Extreme Loading of Structures

Extreme Loading of Structures serves as a valuable resource for graduate studies or as a reference for practicing engineers and covers various topics, including tornado and tornado-generated missiles, vehicular collision, vessel collision, blast, ice load, earthquake ground motion, and more. While focusing mainly on extreme loadings, analytical procedures through which the effects of extreme loads on structures can be assessed are included as well. National design standards and other design specifications are referenced and used throughout the text.

Features:

- Offers comprehensive coverage on extreme loading scenarios such as tornadoes, vehicular and vessel collisions, blasts, ice loads, and earthquake ground motions.
- Provides analytical methods for assessing various load impacts on structures, referencing national design standards and specifications throughout.
- Systematically organizes specific types of extreme load into separate chapters, with detailed explanations of related design criteria and computational procedures for each.

Extreme Loading of Structures

With a Focus on Earthquake Ground Motion

Tim Huff

CRC Press
Taylor & Francis Group
Boca Raton London New York

CRC Press is an imprint of the
Taylor & Francis Group, an **informa** business

Designed cover image: Shutterstock

First edition published 2025
by CRC Press
2385 NW Executive Center Drive, Suite 320, Boca Raton FL 33431

and by CRC Press
4 Park Square, Milton Park, Abingdon, Oxon, OX14 4RN

CRC Press is an imprint of Taylor & Francis Group, LLC

ISBN: 978-1-032-88548-3 (hbk)
ISBN: 978-1-032-88549-0 (pbk)
ISBN: 978-1-003-53836-3 (ebk)

DOI: 10.1201/9781003538363

Typeset in Times
by codeMantra

Access the Support Material: www.routledge.com/9781032885483

Contents

Preface

This text is intended for use by practicing structural engineers and as a text for a graduate-level course in structural engineering. The author has spent the great majority of his 40 years as a structural engineer in engineering for extreme events. The design strategy for extreme loading is quite distinct from that for every day, service loading. Controlled damage is often expected for structures subjected to extreme loadings. Accurate representation of earthquake loading, for example, requires attention to not only the elements that are expected to experience some level of damage but also capacity-protection of those elements expected to remain essentially elastic. While working as a structural engineer with Martin Marietta Energy Systems in Oak Ridge, Tennessee, the author was first introduced to extreme loadings, such as tornadoes, tornado-generated missiles, and earthquakes. This instilled a desire to learn more and apply more of the science of extreme loadings to structural design. While working as a bridge engineer with the Tennessee Department of Transportation, the author continued to focus on extreme loadings, including vehicular impact, vessel collision, and earthquake engineering. This book has grown out of experience acquired over 40 years as a practicing structural engineer. By far the lengthiest chapter is that on earthquake loading, as that has been the greatest focus of a 40-year career which continues in academia at Tennessee Tech University and in consulting work.

Acknowledgments

I am ever grateful for the education I received at Livingston Academy in Livingston, Tennessee. I went on to study Civil Engineering at Tennessee Technological University, receiving a master's in 1985. I'll never forget Dr. Dean Freitag and Dr. Noel Tolbert from the Civil Engineering Department. My master's degree in Mathematics came next, and Dr. Sandra Scheick was instrumental in guiding me at Tennessee State University. My Ph.D. from the University of Tennessee at Knoxville would not have been possible without the influence from Dr. Ed Burdette, Dr. James Mason, and Dr. Richard Bennett.

As a practicing engineer, I have been blessed to work under the direction of Charles T. McLoughlin, Edward P. Wasserman, and Henry Pate. They don't make 'em like these any more.

My family has always encouraged me, through success and failure alike. I am immensely blessed to say that my mother is Marva Sue Story Huff from Byrdstown, Tennessee, and my father is William Cyril Huff from Moodyville, Tennessee. Byrdstown and Moodyville are in Pickett County, the smallest county in Tennessee. My big brother, Troy, and my little sister, Holli, gave me moments of unmeasurable joy in my childhood. My beautiful wife, Monica, is a treasure to me, and I probably would never have written any of my three books without her encouragement. Esteban and Majo are so brave. I think I could learn a lot from them.

About the Author

Tim Huff has more than 40 years of experience as a practicing structural engineer. He has worked on building and bridge projects in the United States and has contributed to projects in India, Ethiopia, Brazil, the Philippines, and Haiti as a volunteer structural engineer with Engineering Ministries International. He is a faculty member of the Civil & Environmental Engineering Department at Tennessee Technological University in Cookeville, where he resides with his beautiful and talented wife, Monica, an artist and teacher.

1 Introduction

1.1 PRELIMINARIES

Extreme loading of structures is generally treated differently in design specifications compared to other limit states, which include service, strength, and fatigue criteria. Under extreme event loading, the structure is often permitted to experience controlled damage and inelastic behavior.

Figure 1.1 illustrates the concept of permitting inelastic behavior for structures subjected to extreme loadings. Service loads are the actual estimated loads due to dead loads, live loads, snow loads, etc. Strength loads are service loads amplified by load factors appropriate for each source. In a properly designed structure, the strength (factored loads) will never be experienced.

Extreme loads include the following:

- Tornado and tornado-generated missiles,
- Vehicular collision,
- Vessel collision,
- Blast loads from explosions,
- Ice loads, and
- Earthquake ground motion.

Each of these is considered in this text. The material is appropriate for practicing engineers as a reference and for graduate students in structural engineering.

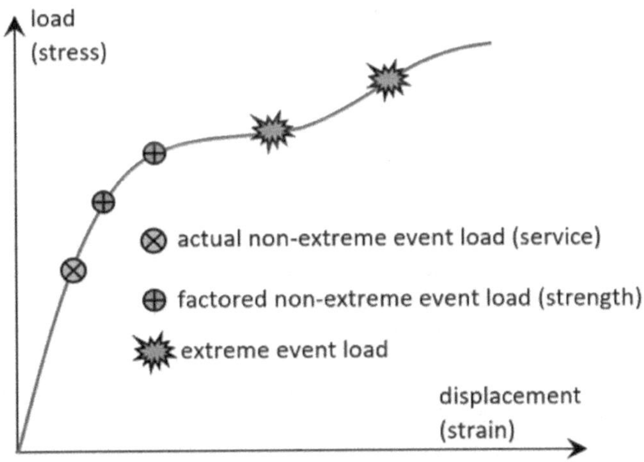

FIGURE 1.1 Extreme versus service/strength structural loading.

DOI: 10.1201/9781003538363-1

The focus of this book is on determining extreme loads, rather than on structural analysis methods for those extreme loads. Nonetheless, some analytical techniques and issues are explored to gain a fuller appreciation of the loadings discussed. Example problems sufficient to illustrate the principles of extreme loading are included.

One important concept in structural design for extreme loads is the idea of "capacity protection". This refers to the identification of the ductile "fuse", with subsequent design and detailing to ensure that all other elements remain essentially elastic. The "fuse" is the weakest element in the path of the resisting system but also the most ductile element. For example, the concept is critical in design for seismic ground shaking. The fuse of the lateral force-resisting system must be identified, designed, and detailed to behave in a ductile manner, while all other elements are designed for a force distribution based on the over-strength of the fuse elements. The fuses actually limit the load effects which can physically be imparted to the protected elements. A few examples, among the many available systems with regard to seismic design, are listed in Table 1.1.

Often, probabilistic features are present in the design of structures for extreme event loading. The geometric probability of collision is of primary importance in the design of bridge structures to withstand vessel collision forces. The probability of exceedance of various ground motion intensity measures is a factor in determining the design seismic loading of structures. The mean recurrence interval of wind speed at a given site is factored into the design of wind pressures for structures.

In structural design, the so-called load factors and resistance factors are typically incorporated into design provisions in modern codes and specifications. Load factors for strength criteria are usually greater than one (as a measure of safety to account for uncertainty in load estimation), while resistance factors are typically less than one (again, to account for uncertainty in the estimation of reliable resistance). For extreme loading, both load and resistance factors may be equal to one in some cases to obtain the best estimates for loading and resistance.

TABLE 1.1
Fuse Elements of Lateral Force-Resisting Systems

System	Fuse Members	Protected Members
Building moment frames	Flexural plastic hinges in beams	Connections and columns
Building concentrically braced frames	Braces yielding in tension and buckling in compression	Beams, columns, and connections
Building eccentrically braced frames	Plastic deformation of links	Beams, columns, braces, and connections
Bridge supports of conventional design	Plastic hinges in the support columns	Cap, superstructure, and foundations
Bridge supports with isolation devices	Bearings	Superstructure, cap, columns, and foundations

There are numerous codified requirements for the extreme loading of structures. Some of these are now discussed in this chapter.

- ASCE 7 (ASCE, 2022): ASCE 7 specifies design loading for extreme events, including earthquake, tsunami, and tornado. The provisions are generally applicable to building systems.
- ASCE 43-19 (ASCE, 2019): ASCE 43 provides seismic design criteria for nuclear safety-related structures.
- ASCE 4-16 (ASCE, 2016): ASCE 4 provides requirements for the seismic analysis of nuclear safety-related structures and is used in conjunction with ASCE 43.
- ASCE 59-11 (ASCE, 2011): ASCE 59 covers performance requirements and loading specifications for the protection of buildings subjected to blast loading.
- AASHTO Guide Specifications for Load and Resistance Factor Design (LRFD) Seismic Bridge Design (AASHTO, 2023a): This guide specification provides guidance on earthquake loading for bridge structures and includes guidance on the evaluation of bridge elements as well.
- AASHTO Guidelines for Performance-Based Seismic Design of Highway Bridges (AASHTO, 2023b): This guideline introduces performance-based design provisions applicable to bridge structures. Performance-based design is becoming more popular for extreme event loading. This guideline specifies structural criteria at two levels of seismic loading.
- AISC N690-18 (AISC, 2018): This standard contains provisions for earthquake, tornado, missile impact, pipe break accidents, and other extreme loading conditions.

These codes, guides, and specifications provide the background for the material presented subsequently.

The following outline will provide the reader with a map of the content provided in this book.

- Chapter 1. Introduction.
- Chapter 2. Tornado and Tornado-Generated Missiles. This chapter presents equations used in determining tornado-based pressure on building surfaces. Various methods of estimating the effect of tornado-generated missile impact are included.
- Chapter 3. Vehicular Collision. This chapter presents material from the American Association of State Highway and Transportation Officials (AASHTO) for the design of bridge barriers subjected to the collision of vehicles.
- Chapter 4. Vessel Collision. This chapter presents material from the American Association of State Highway and Transportation Officials (AASHTO) for the design of bridge substructures subjected to the collision of vessels on navigable waterways.

- Chapter 5. Blast Loading. This chapter presents a discussion of the effects of nearby explosions on buildings. The determination of blast loading and methods used to assess those effects are included.
- Chapter 6. Ice Loading. This chapter provides a brief discussion of methods used to determine the magnitude of ice loads on bridge structures.
- Chapter 7. Earthquake Ground Motion. This chapter is by far the lengthiest chapter in this book. Earthquake engineering has been, and continues to be, a developing science. Topics include the development of design response spectra, the development of design ground motion suites, characterization of earthquake ground motion, ground motion databases, ground motion parameters, artificial ground motions, and design strategies to resist the effects of earthquake ground motion. The material is limited to ground-shaking effects. Other effects, such as liquefaction and lateral spreading, are not covered here.

2 Tornado and Tornado-Generated Missiles

2.1 PRELIMINARIES

ASCE 7-22(ASCE, 2022) contains an entire chapter (Chapter 32) on tornado loads, separate from the wind load provisions in Chapters 26–31. Structures in Risk Categories III and IV are required to resist the greater of the wind loading from Chapters 26–31 or tornado loading from Chapter 32. The focus of this chapter is on the extreme tornado loading on the main wind force-resisting system (MWFRS) from Chapter 32 of ASCE 7-22. The reader is referred to ASCE 7-22 for straight wind loading criteria.

A brief discussion on both risk category designation and exposure category designation is appropriate before covering tornado loading details. The appropriate risk category is an important factor in the magnitude of not only tornado loading but also other extreme loading such as ice and earthquake.

2.2 RISK CATEGORY IN ASCE 7-22

Risk category designation is related to the potential risk to human life should the structure experience failure. Table 2.1 provides basic guidelines for establishing the risk category for a building structure. For detailed further guidance, refer to ASCE 7-22 Chapter 1 Provisions and Chapter 1 Commentary.

TABLE 2.1
Risk Category in ASCE 7-22

Risk Category	Comments	Examples
I	Low risk to human life	Barns, uninhabited storage facilities
II	Not I, III, or IV	Majority of buildings
III	Substantial risk to human life or high economic impact; inhabited by persons with limited ability to escape in a failure event	Theaters, lecture halls, elementary schools, small health-care facilities, prisons, power-generating stations, water treatment plants, some facilities housing explosive, toxic, or otherwise hazardous substances
IV	Essential facilities; substantial hazard to community	Hospitals, police stations, fire stations, emergency communications centers; some facilities housing explosive, toxic, or otherwise hazardous substances

DOI: 10.1201/9781003538363-2

Extreme loading of nuclear safety-related structures is specified in separate standards. ASCE 4-16 (ASCE, 2016) and ASCE 43-19 (ASCE, 2019) are two standards which may supersede ASCE 7-22 with regard to extreme loading criteria.

2.3 WIND EXPOSURE CATEGORIES IN ASCE 7-22

ASCE 7 classifies exposure categories according to the surrounding terrain and obstructions (surface roughness conditions) for a building structure. There are three surface roughness conditions in ASCE 7, summarized in Table 2.2. There are also three exposure categories in ASCE 7, which are summarized in Table 2.3.

2.4 TORNADO WIND PRESSURE

Pressure on building surfaces subjected to tornado loading includes both external pressure and internal pressure. Pressures with a positive sign act toward a given surface, whether internal or external. Pressures with a negative sign act away from a surface, whether internal or external.

Several parameters are necessary to determine the magnitude of tornadic wind pressure. The basic pressure (q_{zT}) equation in Equation 2.1 is used to determine the

TABLE 2.2
Surface Roughness in ASCE 7

Surface Roughness	Description
B	Closely spaced obstructions the size of single-family dwellings or larger
C	Open terrain; scattered obstructions less than 30 ft
D	Flat, unobstructed surfaces of ground or water

TABLE 2.3
Wind Exposure Categories in ASCE 7-22

Exposure Category	Description
B	For buildings with mean roof height no more than 30 ft: Surface Roughness B for a distance greater than 1,500 ft
	For buildings with mean roof height greater than 30 ft: Surface Roughness B for a distance greater than 2,600 ft or 20 times the building height, whichever is larger
C	Not exposure category B or D
D	Surface Roughness D for a distance greater than 5,000 ft or 20 times the building height, whichever is larger.
	Surface Roughness B or C immediately adjacent to the building with Exposure D conditions as defined above within 600 ft or 20 times the building height, whichever is larger, from the building

pressure (p_T) on the main wind force-resisting system (MWFRS) in Equation 2.2. Both q_{zT} and p_T are measured in pounds per square foot (psf).

$$q_{zT} = 0.00256 K_{z\text{Tor}} K_e V_T^2, \tag{2.1}$$

$$p_T = q G_T K_{dT} K_{vT} C_p - q_i \left(G C_{piT} \right), \tag{2.2}$$

where K_e is the ground elevation factor, $K_{z\text{Tor}}$ is the exposure coefficient, V_T is the tornadic wind speed in miles per hour, q: q_{zT} evaluated at the appropriate height above ground, psf, q_i: q for enclosed or partially open buildings, q_i: $q_{z\text{op}}$ for partially enclosed buildings, z_{op} is the level of lowest opening affecting internal pressure, G_T is the gust effect factor, K_{dT} is the directionality factor, K_{vT} is the adjustment for vertical wind effects, C_p is the pressure coefficient, and GC_{piT} is the internal pressure coefficient.

The ground elevation factor is a function of the ground elevation above sea level, z_e, and is given in Equation 2.3. The exposure coefficient at a given elevation depends on the height above ground at that level, z. Equation 2.4 defines $K_{z\text{Tor}}$ for z values in the range of 200–328 ft. For z less than 200 ft, $K_{z\text{Tor}} = 1.0$. For z greater than 328 ft, $K_{z\text{Tor}} = 0.90$.

$$K_e = e^{-0.0000362 z_e}, \tag{2.3}$$

$$K_{z\text{Tor}} = \left(\frac{2{,}820 - z}{2{,}620} \right)^2, \tag{2.4}$$

where z_e is the ground elevation above sea level, ft, and z is the height above ground, ft.

With regard to straight-wind speeds, the gust effect factor is 0.85 for a rigid structure. In ASCE 7-22, a rigid structure is defined as either:

a. a structure with a fundamental, natural frequency no less than 1.00 Hz, or
b. a low-rise structure.

Furthermore, a low-rise structure is defined as a structure with a mean roof height, h:

a. no greater than 60 ft and
b. no greater than the least horizontal dimension of the structure.

For straight-wind speed analysis, complicated expressions are available in ASCE 7-22 for computation of G_T for flexible structures.

With regard to tornadic wind speeds, ASCE 7-22 permits G_T to be computed from Equation 2.5 or simply taken equal to 0.85, regardless of the building frequency. As noted in the Commentary to Section 32.11 of ASCE 7-22, the duration of a tornado is "sufficiently short such that the gust factor provisions of Section 26.11.5 for flexible or dynamically sensitive buildings and other structures do not apply".

$$G_T = 0.925\left(\frac{1+5.78I_{\bar{z}}Q}{1+5.78I_{\bar{z}}}\right), \tag{2.5}$$

$$I_{\bar{z}} = 0.20\left(\frac{33}{\bar{z}}\right)^{1/6}, \tag{2.6}$$

$$Q = \sqrt{\frac{1}{1+0.63\left(\frac{B+h}{L_{\bar{z}}}\right)^{0.63}}}, \tag{2.7}$$

$$L_{\bar{z}} = 500\left(\frac{\bar{z}}{33}\right)^{0.20}, \tag{2.8}$$

$$\bar{z} = 0.6h \geq 15 \text{ ft}. \tag{2.9}$$

B is the horizontal plan dimension perpendicular to the wind direction, ft, and h is the mean roof height, ft.

Figure 2.1 reveals trends in the computed gust effect factor for five different mean roof height values. Given the uncertainties in many of the parameters used to estimate pressures from tornado winds, it is easy to see why a simplistic value of 0.85 is permitted in lieu of the complex equation.

The adjustment factor for vertical wind effects, K_{vT}, is 1.1 for the MWFRS and ranges from 1.0 to 1.3 for components and cladding.

The directionality factor, K_{dT}, is equal to 0.80 for the MWFRS and ranges from 0.75 to 1.0 for components and cladding, depending on the use of the facility.

The internal pressure coefficient, GC_{piT}, for the MWFRS is taken equal to:

- +1.00, for sealed buildings,
- ±0.55, for partially enclosed buildings, and
- 0.00, for open buildings.

For definition of sealed, open, and partially enclosed buildings, refer to Section 26.2 of ASCE 7-22.

For the MWFRS, the pressure coefficient, C_p, is as follows:

- 0.80, windward wall,
- −0.50, leeward wall with $L/B \leq 1.00$,
- −0.30, leeward wall with $L/B = 2.00$,
- −0.20, leeward wall with $L/B \geq 4.00$,
- −0.70, side walls, and
- varies from −0.18 to −1.3 for roofs (see ASCE 7-22 for details).

The tornado speed, V_T, is mapped in ASCE 7-22 and varies in tornado-prone regions of the United States from 50 mph for Risk Category III structures having a small

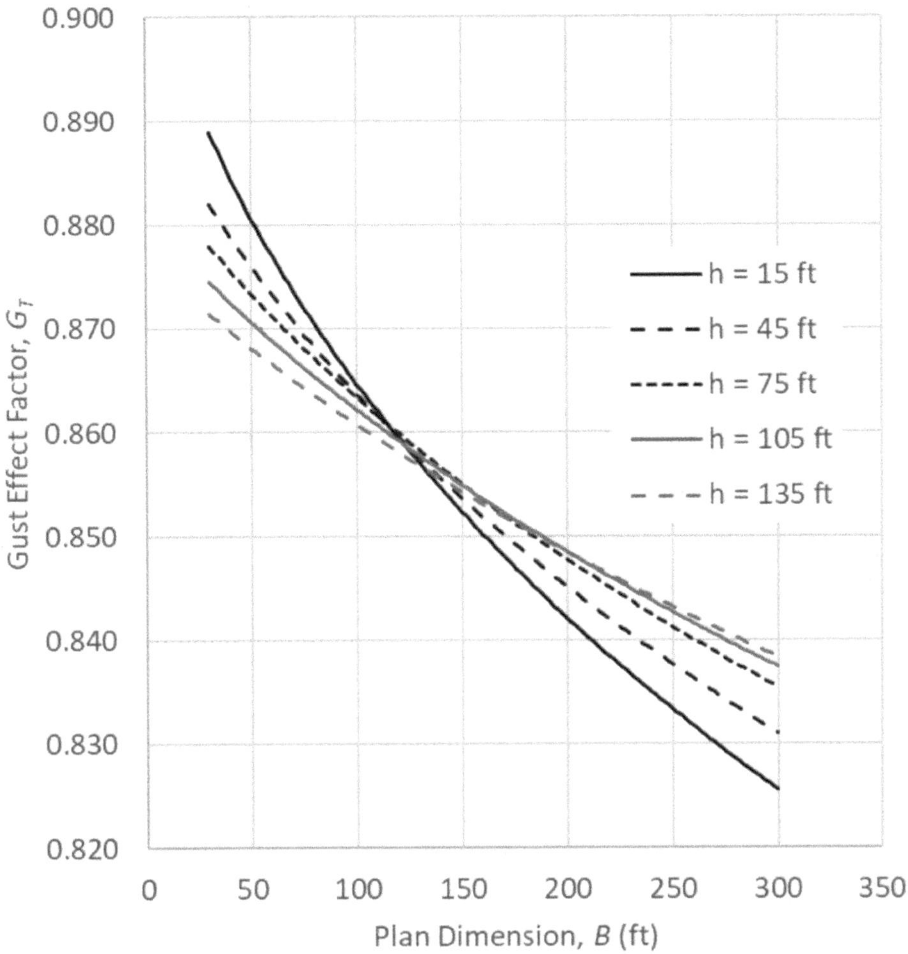

FIGURE 2.1 Gust effect factor for tornado loading.

plan area up to 138 mph for Risk Category IV structures having very large plan area. These are design tornado speeds for the design of conventional buildings in tornado-prone areas.

For the design of safe rooms and tornado shelters, governance is provided by ICC 500: The ICC/NSSA *Standard for the Design and Construction of Storm Shelters*, which references ASCE 7. However, the design tornado speeds in ICC 500 are greater than those in ASCE 7. The intent of ASCE 7 for conventional facilities is to provide life safety, while the intent of ICC 500 for storm shelters is apparently to provide "near absolute protection" for occupants.

For example, while the maximum tornado speed in ASCE 7 is 138 mph for conventional buildings, the maximum tornado speed in ICC 500 is 250 mph for storm shelters in the areas most prone to tornadoes.

ASCE 7-22 does provide an Appendix G, which presents maps for multiple mean recurrence intervals (MRI) ranging from 10,000 to 10,000,000 years. The use of Appendix G is not mandatory but is permitted.

The ASCE 7 Hazard Tool (American Society of Civil Engineers, 2024) provides tornado loading data, as well as data for other extreme loading conditions. The application returns the ASCE life-safety-based tornado speed, as well as the tornado speeds corresponding to the various increasing MRI values. For example, at Cookeville Regional Medical Center in Cookeville, Tennessee (Latitude 36.170°, Longitude −85.509°), Table 2.4 may be produced form the Hazard Tool.

The effective plan area, A_e, is required for interpreting tornado speed values. ASCE 7-22 defines the effective plan area as follows:

> For essential facilities and buildings and other structures required to maintain the functionality of Essential Facilities, the effective plan area shall be equal to the area of the smallest convex polygon enclosing both the Essential Facility and all of the buildings and other structures that maintain the functionality of the Essential Facility.
>
> For buildings and structures that are not designated as Essential Facilities and are not required to maintain the functionality of Essential Facilities, the effective plan area shall be equal to the area of the smallest convex polygon enclosing the plan of the building, other structure, or facility.

The ASCE 7 Hazard Tool notes that the effective plan area should be computed and rounded up to the next available mapped value. Alternatively, log-based interpolation is permitted between tabulated values.

Consider the net lateral pressure on the MWFRS of a 40,000 ft² Risk Category IV building in Cookeville, Tennessee (elevation 1,128 ft; 200 ft or less in height). The windward pressure coefficient and the leeward pressure act in the same direction (toward the windward surface and away from the leeward surface), and the internal pressure on the windward and leeward walls offset one another for the MWFRS (though this is not the case for components and cladding). For these conditions, Equation 2.10 is obtained by setting:

- $K_{zTor} = 1.0$,
- $K_e = 0.96$,

TABLE 2.4
ASCE Hazard Tornado V_T (mph) Results for Cookeville Regional Medical Center

A_e, ft²	3,000 year MRI Risk Cat. IV	10,000 year MRI	100,000 year MRU	1,000,000 year MRI	10,000,000 year MRI
1	84	113	162	205	243
2,000	87	116	164	208	246
10,000	90	119	166	209	248
40,000	93	123	171	212	251
100,000	98	128	174	217	254
250,000	103	133	180	220	258
1,000,000	118	143	189	229	264
4,000,000	130	155	199	238	274

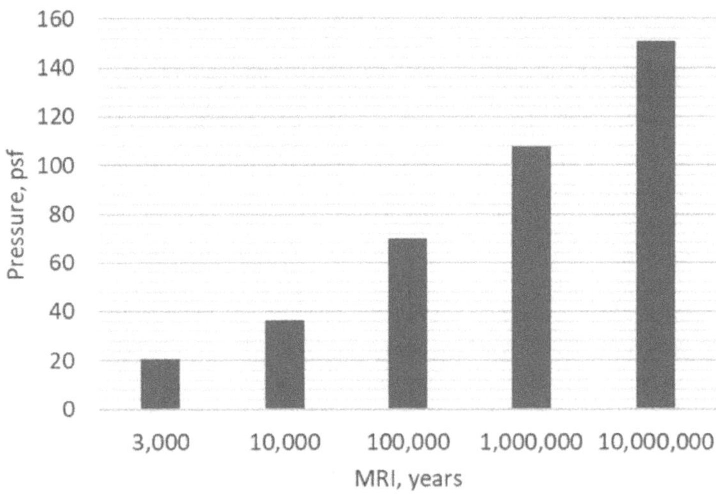

FIGURE 2.2 Net MWFRS pressure in Cookeville, Tennessee.

- $G_T = 0.85$,
- $K_{dT} = 0.80$,
- $K_{vT} = 1.1$,
- $C_p = 1.3$ (0.80, windward and 0.50, leeward),

$$q_{zT} = 0.00239 V_T^2. \tag{2.10}$$

Figure 2.2 shows the wind pressures resulting from the wind speeds tabulated in Table 2.4. The net pressure on the MWFRS for tornado loading is a challenging load to accommodate, both for the MWFRS and for components and cladding, where localized effects and the combination of external and internal pressures do not always offset each other.

2.5 TORNADO-GENERATED MISSILES

Particularly for structures housing hazardous and toxic substances, it may be necessary to design components and cladding to resist the impact from tornado-generated missiles.

Department of Energy facilities have included missile impact criteria over the years in various documents (Kennedy, et al., 1990; Singhal & Walls, 1993). For high-hazard facilities, some tornado-generated missile loadings that have been specified include the following:

- a 2-inch× 4-inch, 15-lb timber plank at 150 mph horizontal velocity or 100 mph vertical velocity at a maximum height of 200 ft,
- a 3-inch standard pipe weighing 75 lbs at 75 mph horizontal velocity or 50 mph vertical velocity at a maximum height of 100 ft, and
- a 3,000-lb rolling automobile at 25 mph horizontal velocity.

TABLE 2.5
Threshold Perforation Velocities – Timber Plank

Barrier	Perforation V (mph)
Masonite siding	54
Plywood, ½ inch	52
Plywood, ¾ inch	53
Stucco	53
Unreinforced CMU wall, 8 inches thick	60
8-inch CMU, grout and rebar in every cell	>130
8-inch CMU, horizontal reinforcement only	65
Unreinforced CMU wall, 12 inches thick	65
12-inch CMU, grout and rebar in every cell	>130
12-inch CMU, horizontal reinforcement only	70

Singhal and Walls (1993) reported threshold perforation velocities for construction materials impacted by a 2-inch×4-inch timber plank weighing 15 lbs. The threshold velocity is the velocity at which the missile just passed through the thickness of the material. Table 2.5 summarizes the threshold perforation velocities for various materials impacted by the plank. Concrete masonry unit (block) wall barriers are abbreviated as CMU in the table. In the same report, equations for steel and concrete barriers are provided and given here in Equations 2.11 and 2.12.

$$T_p = \frac{\left(0.5MV_s^2\right)^{2/3}}{672d_o}, \tag{2.11}$$

$$x = \begin{cases} \sqrt{4KNW_m d_o\left(\dfrac{V_s}{1{,}000d_o}\right)^{1.8}} & \text{for } x/d_o \le 2 \\ d_o + KNW_m\left(\dfrac{V_s}{1{,}000d_o}\right)^{1.8} & \text{for } x/d_o > 2 \end{cases}. \tag{2.12}$$

The report by Singhal and Walls recommends a 25% increase in the calculated value from Equation 2.11 for steel barriers and a 20% increase in the calculated value from Equation 2.12 for concrete barriers.

where d_o is the nominal diameter, inches; f_c' is the concrete compressive strength, psi; K is the concrete penetrability factor $= 180/(f_c')^{0.5}$; M is the missile mass, lb-s²/ft; N is the missile shape factor $= 0.72$ for flat-nosed missiles; t_p is the steel barrier thickness required to prevent penetration $= 1.25T_p$, inches; T_p is the steel barrier thickness just perforated, inches; V_s is the striking velocity of missile, *ft*/s; W_m is the missile weight, pounds; and x is the penetration depth into concrete, inches.

TABLE 2.6
Minimum Barrier Thickness for Tornado-Generated Missile Protection

Region	f_c' MPa (psi)	Wall Thickness cm (Inches)	Roof Thickness cm (Inches)
I	20.7 (3,000)	46.2 (18.2)	33.5 (13.2)
I	27.6 (4,000)	42.9 (16.9)	31.2 (12.3)
I	34.5 (5,000)	40.6 (16.0)	29.7 (11.7)
II	20.7 (3,000)	39.1 (15.4)	28.4 (11.2)
II	27.6 (4,000)	36.3 (14.3)	26.4 (10.4)
II	34.5 (5,000)	34.5 (13.6)	25.1 (9.9)
III	20.7 (3,000)	30.2 (11.9)	22.1 (8.7)
III	27.6 (4,000)	28.2 (11.1)	20.6 (8.1)
III	34.5 (5,000)	26.7 (10.5)	19.6 (7.7)

Using Equation 2.11, it may be found that a 0.29-inch-thick steel plate will just be perforated by the 75-lb, 3-inch diameter steel pipe. The recommended barrier thickness to prevent such perforation is thus $1.25 \times 0.29 = 0.36$ inches (3/8-inch-thick plate).

Using Equation 2.12, it may be found that a 2.47-inch-thick 2,500 psi concrete barrier will just be perforated by the 75-lb, 3-inch diameter steel pipe. The recommended barrier thickness to prevent such perforation is thus $1.20 \times 2.47 = 2.96$ inches (3-inch-thick concrete).

The Nuclear Regulatory Commission (U.S. Nuclear Regulatory Commission, 2007; United States Atomic Energy Commission, 1974) specifies minimum acceptable concrete barrier thickness requirements for three separate regions of the continental United States. Region I is generally east of the Rocky Mountains and is the most prone to tornado loading of the three regions. Region II includes the Pacific coast of the United States. Region III is composed of the remainder of the continental United States. Table 2.6 summarizes the minimum requirements to prevent perforation, spalling, or scabbing from tornado-generated missiles. Smaller barrier thicknesses may be used if validated by testing, analysis, or a combination of testing and analysis.

3 Vehicular Collision

AASHTO (2020) Section 3.6.5 defines vehicular collision loads for the design of piers and abutments that may experience such loads. Design choices include the following:

- providing adequate structural resistance of the abutment or pier elements and
- redirection or absorption of the potential collision load.

To provide structural resistance, AASHTO specifies a 600-kip static load applied at an angle of 0°to 15° with the pavement edge, and at a distance from 2 to 5 ft above the ground. The angle and distance for design are those producing the most severe shear or moment in the element under investigation.

For bridge parapet design, Chapter 13 of the AASHTO Specifications defines several test levels (TL) applicable to various conditions. Table 3.1 summarizes the test levels for barriers. The parameters are defined as follows:

- F_t is a transverse vehicular impact force distributed over a length, L_t, at a height, H_e, above a bridge deck.
- F_L is a longitudinal force equal to 0.33 F_t and distributed over a length, L_L.
- F_v is a vertical force due to a vehicle laying on top of a barrier and distributed over a barrier length equal to L_v.

Test-level 1 (TL-1) criteria apply for work zones with low posted speeds and for very low-volume local routes and streets.

Test-level 4 (TL-4) criteria are generally acceptable for the majority of high-speed highways, freeways, and interstates with a mixture of trucks and heavy vehicles.

TABLE 3.1
AASHTO Test-Level Specifications for Vehicular Collision

Parameter	TL-1	TL-2	TL-3	TL-4	TL-5	TL-6
F_t, transverse (kips)	13.5	27.0	54.0	54.0	124.0	175.0
F_L, longitudinal (kips)	4.5	9.0	18.0	18.0	41.0	58.0
F_v, vertical (kips)	4.5	4.5	4.5	18.0	80.0	80.0
L_t and L_L (ft)	4.0	4.0	4.0	3.5	8.0	8.0
L_v (ft)	18.0	18.0	18.0	18.0	40.0	40.0
H_e (min., inches)	18.0	20.0	24.0	32.0	42.0	56.0
Min rail height, H (inches)	27.0	27.0	27.0	32.0	42.0	90.0

DOI: 10.1201/9781003538363-3

Test-level 6 (TL-6) criteria are for applications involving tanker-type truck or other high-center-of-gravity vehicles, where unfavorable site conditions exist.

For other test-level discussions, refer to Section 13: Railings of the AASHTO LRFD Bridge Design Specifications (AASHTO, 2020), also known as the AASHTO LRFD-BDS.

For bridge parapet design, the parapet resistance may be computed from yield-line theoretical considerations. The AASHTO LRFD-BDS contains equations that permit the engineer to estimate parapet resistance, R_W, based on yield-line theory. Separate equations are provided for interior parapet segments and end parapet segments.

- H is the actual height of the wall or barrier, in ft.
- M_c is the flexural resistance of the barrier or wall about an axis parallel to the longitudinal axis of the bridge, in ft-kips per ft.
- M_w is the flexural resistance of the wall or barrier about a vertical axis, in ft-kips.
- M_b is the added flexural resistance of any beam attached to the top of a barrier or wall about a vertical axis, in ft-kips.
- R_w is the total transverse resistance of the wall or barrier from yield-line theory, in kips.

AASHTO does note that the equations for R_w are applicable as long as M_c and M_w do not vary significantly over the height of the wall. If this is not the case, then a rigorous yield-line analysis is required.

For impact within a parapet segment:

$$L_c = \frac{L_t}{2} + \sqrt{\left(\frac{L_t}{2}\right)^2 + \frac{8H(M_b + M_w)}{M_c}}, \tag{3.1}$$

$$R_W = \left(\frac{2}{2L_c - L_t}\right)\left(8M_b + 8M_w + \frac{M_c L_c^2}{H}\right). \tag{3.2}$$

For impact at the end of a parapet:

$$L_c = \frac{L_t}{2} + \sqrt{\left(\frac{L_t}{2}\right)^2 + \frac{H(M_b + M_w)}{M_c}}, \tag{3.3}$$

$$R_W = \left(\frac{2}{2L_c - L_t}\right)\left(M_b + M_w + \frac{M_c L_c^2}{H}\right). \tag{3.4}$$

These equations are applicable to a barrier on a bridge deck and may also be applicable to otherwise supported barriers.

Figure 3.1, taken from the LRFD Bridge Design Specifications (AASHTO, 2020), depicts an interior parapet segment and defines the distances L_c and L_t, as well as

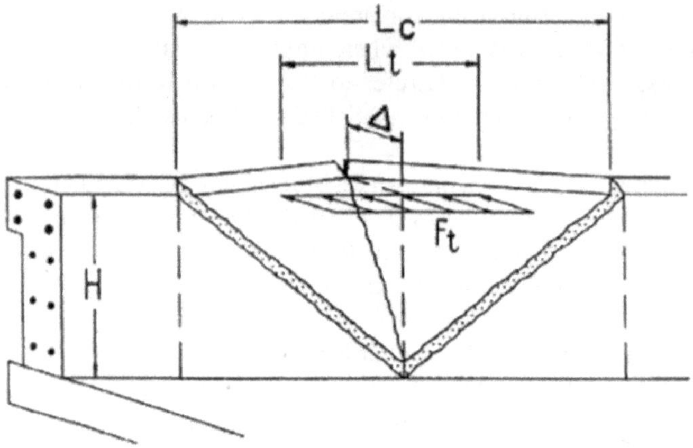

FIGURE 3.1 Vehicular collision – interior segment of a parapet (AASHTO, 2020).

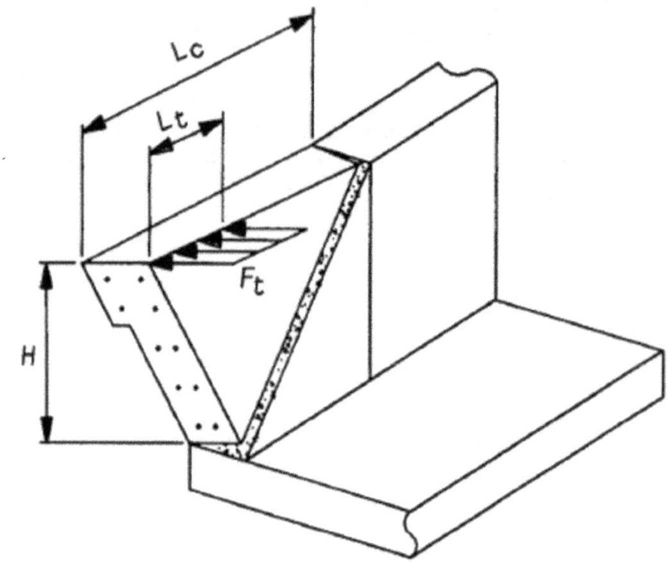

FIGURE 3.2 Vehicular collision – end segment of a parapet (AASHTO, 2020).

the presumed yield-line surfaces. Figure 3.2 is the corresponding graphic for an end segment of parapet.

Consider Figure 3.3, taken from standard drawings of the Tennessee Department of Transportation at: https://www.tn.gov/content/tn/tdot/structures-/standard-structures-drawings/new-structures.html.

The parapet also typically includes a metal rail beam, which has been conservatively neglected in the sample analysis presented here. A sidewalk thickness, T_W, has been taken as zero for this example, which is also a conservative condition. Suppose the parapet is to be used at a location where TL-3 criteria are applicable.

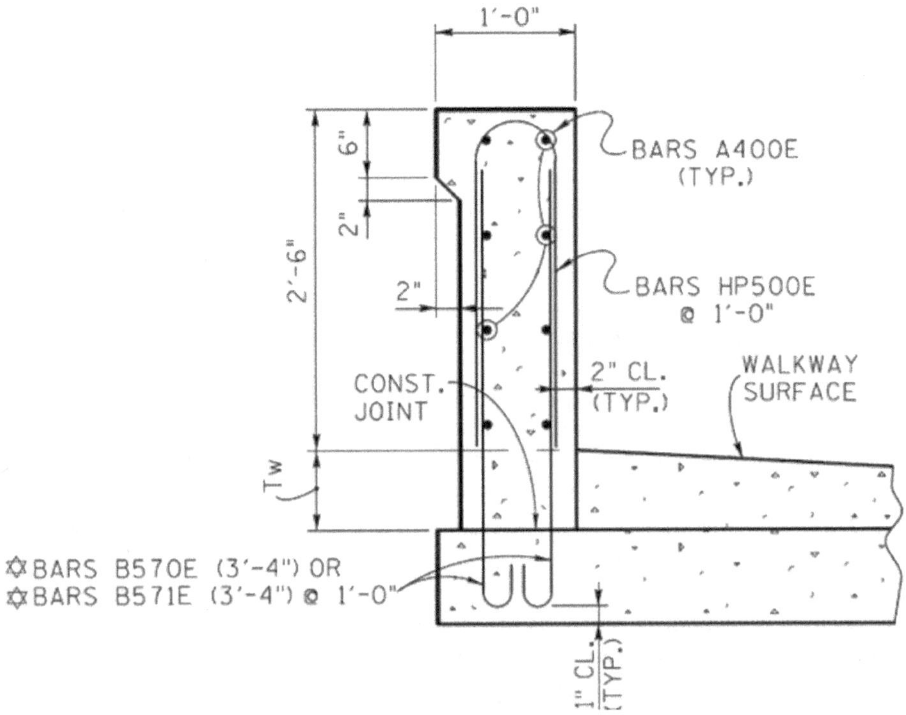

FIGURE 3.3 Tennessee DOT STD-11-1 bridge parapet cross-section.

Cross-section analysis can be used to show that, with the reinforcement shown, parapet capacities are as follows:

- $M_W = 42.9$ ft-kips,
- $M_C = 16.6$ ft-kips per ft.

For TL-3 criteria:

- $L_t = 4$ ft,
- $F_t = 54.0$ kips.

With these parameters, the parapet resistance equations give $R_W = 126$ kips for an interior parapet segment and $R_W = 70$ kips for an end parapet segment. Both values exceed the required resistance of 54.0 kips. The parapet height of 30 inches exceeds the tabulated minimum of 27 inches. Therefore, the parapet satisfies TL-3 criteria.

For barrier design to protect bridge piers, AASHTO defines four scenarios that may be assumed to provide adequate resistance and are not required to be analyzed in detail. These are as follows:

1. Substructure components backed by soil. A great majority of typical bridge abutments fall into this category.

2. Reinforced concrete pier components that are at least 3.0-ft thick and have a concrete cross-sectional area greater than 30.0 ft^2, as measured in the horizontal plane at all elevations from the top of the pier foundation to a height of at least 5.0 ft above the grade.
3. Piers supporting a bridge superstructure where it is shown by calculation that the superstructure will not collapse with one column missing when subjected to the full dead load with a 1.1 load factor, and the live load, including dynamic load allowance, in the permanent travel lanes with a load factor of 1.0.
4. Pier walls and multicolumn piers with struts between columns that have been designed and detailed as Manual for Assessing Safety Hardware (MASH) Test-Level 5 (TL-5) longitudinal traffic barriers according to Section 13.

While a circular column would need to be approximately 75 inches in diameter to have the required area of 30 ft^2 to qualify for the no-analysis criterion, it is possible to carry the 600-kip static load in shear on a single-loaded column with much smaller diameters. Consider, for example, a 16-ft high column acting as part of a rigid-frame pier. Assume further that the base of the column is 1-ft below ground. So, an estimate of the shear produced by the 600-kip load would be $600 \times (13/16) = 487$ kips. While a detailed analysis for flexure-shear interaction would be required, it can be shown that a circular column as small as 46 inches in diameter, with 2.5% longitudinal reinforcement and tightly spaced transverse reinforcement in the form of hoops or spirals, can resist such a shear without unreasonable concrete strength requirements. The collision load is not necessarily prohibitive. Nonetheless, given the cost of bridge construction, pier protection may be the more prudent option when vehicular collision load design is required.

When the design choice is to redirect or absorb the vehicular collision load rather than to design the pier for the impact, the protection must consist of a minimum 42-inch-tall Manual for Assessing Safety Hardware (MASH) crash-tested rigid TL-5 barrier. This barrier must be located such that the top edge of the barrier face on the traffic side is no less than 3.25 ft from the face of the component being protected.

4 Vessel Collision

Section 3.14 of the AASHTO LRFD Bridge Design Specifications (AASHTO, 2020) covers loading due to the collision of vessels into bridge substructures. Design for vessel collision is required for all navigable waterways with water depth not less than 2 ft.

The minimum required impact load for substructure design is that from an empty hopper barge, measuring 35-ft × 195-ft and weighting 200 tons (1 ton = 2,000 lbs). The barge velocity for this minimum required loading is taken equal to the yearly mean current of the waterway. Piers in navigable waterways beyond three times the length overall (LOA) of the design vessel from the centerline of the vessel transit path are designed for this minimum loading. Piers closer to the centerline of the transit path may need to be designed for larger forces.

When the superstructure is subject to potential vessel collision, the minimum design load for the superstructure is based on a mast collision load as specified in Section 3.14.10.3 of the Specifications.

Design for vessel collision, where required, or the protection of bridge substructures by sacrifice-able structures, or a combination of the two may be used to protect bridge piers.

For a particular pier, the design vessel is selected such that the annual frequency of collapse due to collision of vessels not smaller than the design value is less than a prescribed acceptance criterion.

Deadweight tonnage (DWT) is a factor in vessel collision force calculation and includes the sum of cargo, fuel, fresh water, ballast water, provisions, passengers, and crew. DWT does not include the deadweight of the vessel. The displacement weight tonnage, W, on the other hand, includes the DWT plus the self-weight of the vessel.

The head-on impact force from vessel collision on a bridge pier by a ship is given in Equation 4.1. This force is an equivalent static force and is a function of the impact velocity, V (ft/s), and the vessel DWT (tonnes, 1.0 tonne = 2,205 lbs). Barge weight is measured in tons (1 ton = 2,000 lbs), while ship weight is measured in tons. However, AASHTO equations are typically tonne-based.

$$P_S = 8.15V\sqrt{\text{DWT}}. \tag{4.1}$$

The impact force from vessel collision on a bridge pier by a barge, P_B (kips), is given in Equation 4.2. The barge bow damage length, a_B (ft), is from Equation 4.3. The kinetic energy, KE (kip-ft), is given in Equation 4.4.

$$P_B = \begin{cases} 4{,}112a_B & \text{if } a_B < 0.34 \\ 1{,}349 + 110a_B & \text{if } a_B \geq 0.34 \end{cases}, \tag{4.2}$$

$$a_B = 10.2\left(\sqrt{1 + \frac{\text{KE}}{5{,}672}} - 1\right), \tag{4.3}$$

DOI: 10.1201/9781003538363-4

$$\mathrm{KE} = C_H \frac{WV^2}{29.2}. \tag{4.4}$$

For the kinetic energy calculation in Equation 4.4, W is the vessel displacement tonnage (tonne), V is the impact velocity (ft/s), and C_H is a hydrodynamic mass coefficient. The vessel displacement tonnage, W(tonness), includes the empty weight of the vessel plus the weight of cargo for loaded vessels.

If the under-keel clearance (the distance from the bottom of the keel to the seabed) is less than 10% of the draft (the distance from the bottom of the keel to the waterline), then $C_H = 1.05$. If the under-keel clearance is greater than 50% of the draft, then $C_H = 1.25$. Linear interpolation is used for intermediate under-keel clearance values.

For barge collisions, AASHTO specifies a standard hopper barge with the following properties.

- width = 35 ft,
- length = 195 ft,
- depth = 12 ft,
- empty draft = 1.7 ft,
- loaded draft = 8.7 ft, and
- DWT = 1,700 tons.

For the extreme event limit state, inelastic behavior is permitted in the AASHTO Specifications. However, collapse of the structure must still be avoided, even when inelastic behavior forms the basis of design for extreme event loading.

For substructure design, the computed static force is applied under two separate conditions:

1. the full computed impact force in a direction parallel to the channel centerline, and
2. one-half of the computed force in a direction normal to the channel centerline.

For overall stability, the collision force is applied as a concentrated force at the waterline, corresponding to the mean high-water level.

For local collision forces, the application depends on whether the collision is from a ship or a barge. For a ship vessel collision, the force is applied as a distributed load along the bow depth. For a barge collision, the vessel collision force is applied as a distributed load along the head block depth. The head block depth of a barge is typically much shallower than the bow depth of a ship, making barge local collision forces much closer to a concentrated force.

The collision impact velocity for design purposes, V (ft/s), is determined from Equation 4.5. Other variables in the equation are as follows:

- V_T = the typical transit velocity, not to be taken less than V_{MIN} (ft/s),
- V_{MIN} = the minimum impact velocity; not to be taken less than the mean current velocity (ft/s),

- X = distance from pier face to centerline of channel (ft),
- X_C = distance to edge of channel (ft), and
- X_L = three times the overall length of the design vessel (ft).

$$V = \begin{cases} V_T, & \text{if } X \leq X_C \\ V_{MIN}, & \text{if } X \geq X_L \\ V_T - (V_T - V_{MIN})\dfrac{X - X_C}{X_L - X_C}, & \text{if } X_C < X < X_L \end{cases} \quad . \tag{4.5}$$

The vessel length overall, LOA, is used in determining the probability of collapse due to vessel collision forces for a bridge. The annual frequency of bridge collapse (AF) is determined for a trial design vessel. The AF is then compared to acceptance criteria, and revised vessel characteristics are used until a convergence is achieved. The annual frequency of collapse is computed for each element (typically piers) exposed to potential collision forces.

The criteria include the following:

- AF must be computed for each component susceptible to potential collision.
- The AF for the bridge is the sum of each component AF.
- For critical or essential bridges, the bridge AF must be no larger than 0.0001.
- For typical bridges, the AF must be no more than 0.001.

The distribution of required AF values depends on the waterway width. For waterway width less than six times the LOA, the required acceptance criterion is distributed equally across all piers. For waterway widths greater than six times the LOA, the required acceptance criterion is distributed equally to those piers within a distance equal to three times the LOA on each side of the transit path. However, this is not the only possible strategy, as owners may wish to assign more stringent acceptance criteria to main span piers, which are typically of higher cost relative to other substructures. The more stringent criteria would potentially be accompanied by somewhat relaxed criteria for piers other than the main span piers. Regardless, the sum of all component AF values must not exceed 0.0001 for essential and critical bridges, 0.001 for typical bridges.

The annual frequency of collapse due to vessel collision for a bridge component is given in Equation 4.6. The AF depends on five other factors defined below and in Equations 4.7 through 4.11.

- N = the number using the channel annually, classified by type, size, and load,
- PA = the probability of vessel aberrancy,
- PG = the geometric probability of collision,
- PC = the probability of collapse due to an aberrant vessel,
- PF = adjustment factor accounting for potential use of pier protection devices,
- H = resistance of a bridge component, kips, and
- P = vessel impact force, kips.

$$AF = N(PA)(PG)(PC)(PF), \tag{4.6}$$

$$PC = \begin{cases} 0.1 + 9\left(0.1 - \frac{H}{P}\right) & \text{if } 0.0 \le H/P < 0.10 \\ 0.111\left(1 - \frac{H}{P}\right) & \text{if } 0.10 \le H/P < 1.00 \\ 0.00 & \text{if } H/P \ge 1.00 \end{cases}, \tag{4.7}$$

$$PF = 1 - \frac{\text{percent protection provided}}{100}. \tag{4.8}$$

With no pier protection, PF = 1.00. With 100% pier protection, PF = 0.0. PF may vary from pier to pier.

For guidance on the probability of vessel aberrancy PA and the geometric probability of collision PG refer to Section 3.14 of the ASHTO LRFD Bridge Design Specifications (AASHTO, 2020). Detailed, site-specific treatment of these parameters is beyond the scope of this book.

Factors that enter into rigorous calculation of PA include long-term vessel accident data and frequency of ship and barge traffic for the waterway. The AASHTO Specifications provide a simplified approximate method for determining PA when such data do are unavailable. Using current velocities of seven knots parallel to the vessel path and three knots perpendicular to the vessel path, a 90° bend in the waterway, a high density of vessels, and an aberrancy base rate of 0.00012 produces a value of PA equal to 0.0039 (seemingly, an extremely high value) using the approximate method. This sample calculation was based on largely conservative values for the parameters involved. In a study based on inland waterways in Kentucky (Whitney et al., 1996), and specifically the cable-stayed bridge over the Ohio River in Maysville, a value of PA equal to 0.00017 based on site-specific data was used. The average PA for the Ohio River was found to be 0.000529.

The approximate method of computing PA is based on the aberrancy base rate (BR), a correction factor for bridge location (R_B), a correction factor for current acting parallel to the transit path of the vessel (R_C), a correction factor for cross-currents (R_{XC}), and a correction factor for vessel traffic density (R_D). These values are determined as follows and the approximate PA is simply the product of these five factors.

- BR = 0.00006 for ships,
- BR = 0.00012 for barges,
- $R_B = 1.0$ for straight regions,
- $R_B = 1 + \theta/90$ for transition regions, where θ is the angle of turn in degrees,
- $R_B = 1 + \theta/45$ for turn and bend regions,
- $R_C = 1 + V_C/10$, where V_C is the current velocity parallel to the transit path in knots,
- $R_{XC} = 1 + V_{XC}$, where V_{XC} is the cross-current velocity in knots,

- $R_D = 1.0$ in a location of low vessel traffic density,
- $R_D = 1.3$ in a location of average vessel traffic density, and
- $R_D = 1.6$ in a location of high vessel traffic density.

Low, average, and high-density conditions exist where vessels rarely, occasionally, and frequently meet, pass, or overtake in the vicinity of the bridge, respectively.

A "turn region" is an abrupt directional change in the waterway. A bridge is considered to be in a turn region if it is within 3,000 ft of the abrupt change. A bridge located between 3,000 and 6,000 ft from an abrupt turn is considered to be in a transition region. A bridge located more than 6,000 ft from an abrupt turn is considered to be in a straight region.

A "bend" region is a smooth, curved directional change in the waterway. A bridge is considered to be located in a bend region if it lies within the bend. It is considered to be in a "transition" region if it lies within 3,000 ft of the onset of the bend. A bridge which lies farther than 3,000 ft from the onset of a bend is considered to be in a straight region.

Factors that enter into the calculation of the geometric probability, PG, include waterway geometry, pier locations, span clearances, vessel path and geometry, environmental conditions, and vessel draft, among others. In the study mentioned above (Whitney et al., 1996) on the discussion of PA, the authors adopted PG values ranging from 0.034 to 0.119 for barge traffic on the Ohio River.

The geometric probability, PG, is defined as the "area under the normal distribution bounded by the pier width and the width of the vessel on each side of the pier". Presumably, the pier width is the projected width, as shown in Figure 4.1. The bounds of the shaded area are $X - (B_P + B_M)/2$ and $X + (B_P + B_M)/2$. With X taken equal to the

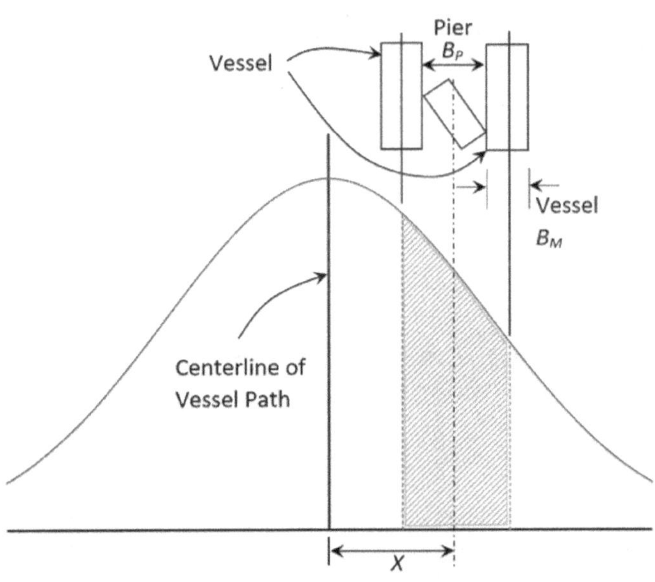

FIGURE 4.1 Geometric probability calculation for vessel collision with a bridge pier.

actual distance from the vessel path centerline divided by the vessel length, LOA, the standard deviation becomes 1, with mean 0. So, each pier has a specific value of PG for each vessel considered.

Note that if the piers are all designed to resist the collision load elastically (i.e., *H*/*P* not less than 1.0), then PC=0 and AF=0. This would likely be cost-prohibitive in many cases where navigable waterways are subject to vessel collision forces.

Examples of costly vessel collision damage to bridge piers, leading to the collapse of at least one span, include the following:

- Interstate 40 over the Arkansas River in 2002,
- Sunshine Skyway Bridge in 1980,
- Popps Ferry Bridge in 2009, and
- Francis Scott Key Bridge in 2024.

Additional references providing valuable insight in the design of bridges for vessel collision include studies by Liu and Wang (2001) and Whitney et al. (1996).

4.1 ILLUSTRATIVE EXAMPLE

The bridge shown in Figure 4.2 is designed for barge traffic. For this simplified example, the centerline of the barge traffic path is 797.5 ft from the beginning of the bridge. The bridge is in a straight segment of the waterway with no cross-current and low traffic density. The pier width is 11 ft. The current velocity is seven knots parallel to the traffic path. The barge transit velocity is ten knots at the bridge site. The under-keel clearance is greater than 50% of the draft at all pier locations. No superstructure elements are subject to collision loading. The anticipated barge traffic is 10 vessels/week. The displacement weight of the design vessel is 1,890 tons (4,167 kips), with a length of 195 ft and a width of 35 ft. The bridge is a typical bridge (as opposed to essential or critical), so the upper limit on the annual frequency of collapse of the system, AF, is 0.001.

Step 1. Establish the probability of vessel aberrancy, PA.

$$\text{PA} = \text{BR}(R_B)(R_C)(R_{CX})(R_D) = 0.00012(1.00)(1+7/10)(1.00)(1.00) = 0.000204.$$

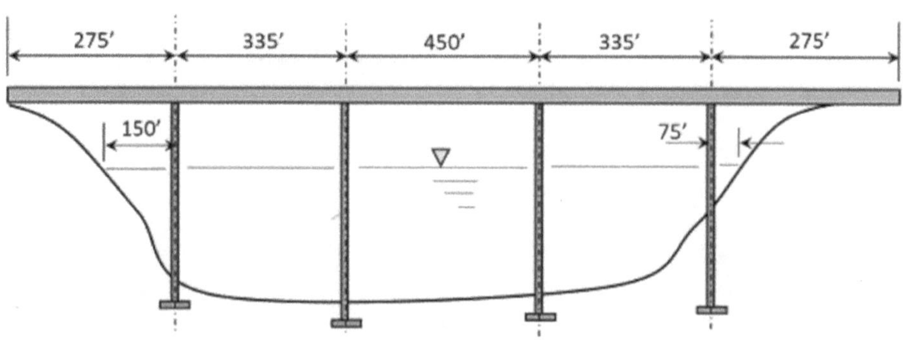

FIGURE 4.2 Vessel collision example bridge.

Step 2. Determine the distances, X_C (the distance from each pier to the edge of the channel) and X_L (three times the vessel length).

$$X_L = 3 \times 195 = 585 \text{ ft},$$

$$X_{C1} = 150 \text{ ft, Pier 1},$$

$$X_{C2} = 150 + 335 = 485 \text{ ft, Pier 2},$$

$$X_{C3} = 75 + 335 = 410 \text{ ft, Pier 3},$$

$$X_{C4} = 75 \text{ ft, Pier 4}.$$

Step 3. Determine X, the distance from each pier to the channel centerline, for each pier.

$$X_1 = 797.5 - 275 = 522.5 \text{ ft, Pier 1},$$

$$X_2 = 522.5 - 335 = 187.5 \text{ ft, Pier 2},$$

$$X_3 = 450 - 187.5 = 262.5 \text{ ft, Pier 4},$$

$$X_4 = 262.5 + 335 = 597.5 \text{ ft, Pier 4}.$$

Step 4. Determine the design velocity of impact, V, at each pier.

$$V_{MIN} = 7 \text{ knots} \times 1.688 \frac{\text{fps}}{\text{knot}} = 11.82 \text{ ft/s},$$

$$V_T = 10 \text{ knots} \times 1.688 \frac{\text{fps}}{\text{knot}} = 16.88 \text{ ft/s}.$$

For Pier 1, note that $X_C<X<X_L$, so interpolation is required to find the velocity, V. A tabulated summary of design velocities is provided below.

Pier	X, ft	X_C, ft	Condition	V, ft/s
1	522.5	150	$X_C<X<X_L$	12.55
2	187.5	485	$X<X_C$	16.88
3	262.5	410	$X<X_C$	16.88
4	597.5	75	$X>X_L$	11.82

Step 5. Determine the vessel kinetic energy, KE.

$$C_H = 1.25.$$

For Pier 1:

$$\text{KE} = C_H \frac{WV^2}{29.2} = 1.25 \frac{1{,}890(12.55)^2}{29.2} = 12{,}743 \text{ ft-kips.}$$

For Pier 2 and Pier 3:

$$\text{KE} = C_H \frac{WV^2}{29.2} = 1.25 \frac{1{,}890(16.88)^2}{29.2} = 23{,}053 \text{ ft-kips.}$$

For Pier 4:

$$\text{KE} = C_H \frac{WV^2}{29.2} = 1.25 \frac{1{,}890(11.82)^2}{29.2} = 11{,}304 \text{ ft-kips.}$$

Step 6. Determine the barge loading coefficient, a_B.
For Pier 1:

$$a_B = 10.2\left(\sqrt{1+\frac{\text{KE}}{5{,}672}} - 1\right) = 10.2\left(\sqrt{1+\frac{12{,}743}{5{,}672}} - 1\right) = 8.18.$$

For Pier 2 and Pier 3:

$$a_B = 10.2\left(\sqrt{1+\frac{\text{KE}}{5{,}672}} - 1\right) = 10.2\left(\sqrt{1+\frac{23{,}053}{5{,}672}} - 1\right) = 12.75.$$

For Pier 4:

$$a_B = 10.2\left(\sqrt{1+\frac{\text{KE}}{5{,}672}} - 1\right) = 10.2\left(\sqrt{1+\frac{11{,}304}{5{,}672}} - 1\right) = 7.45.$$

Step 7. Determine the design collision load at each pier. Note that a_B is greater than 0.34 in all cases.
For Pier 1:

$$P_B = 1{,}349 + 110a_B = 1{,}349 + 110(8.18) = 2{,}248 \text{ kips.}$$

For Pier 2 and Pier 3:

$$P_B = 1{,}349 + 110a_B = 1{,}349 + 110(12.75) = 2{,}752 \text{ kips.}$$

For Pier 4:

$$P_B = 1{,}349 + 110a_B = 1{,}349 + 110(7.45) = 2{,}169 \text{ kips.}$$

Step 8. Determine the geometric probability of collision at each pier.

For Pier 1: $B_P = 11$ ft, $B_M = 35$ ft, $X = 522.5$ ft, $\sigma = \text{LOA} = 195$ ft

$$x_{\max} = \frac{522.5 + 0.5(11 + 35)}{195} = 2.797,$$

$$x_{\min} = \frac{522.5 - 0.5(11 + 35)}{195} = 2.562.$$

Find the area between 2.562 and 2.797 on the normal distribution curve having a mean at zero and a standard deviation of 1. Using statistical tables or Microsoft Excel with the excel function NORM.DIST, find the following:

$$\text{PG}_1 = 0.0213.$$

For Pier 2: $B_P = 11$ ft, $B_M = 35$ ft, $X = 187.5$ ft, $\sigma = \text{LOA} = 195$ ft

$$x_{\max} = \frac{187.5 + 0.5(11 + 35)}{195} = 1.0795,$$

$$x_{\min} = \frac{187.5 - 0.5(11 + 35)}{195} = 0.8436.$$

Find the area between 1.079 and 0.8436 on the normal distribution curve having a mean at zero and a standard deviation of 1. Using statistical tables or Microsoft Excel with the excel function NORM.DIST, find the following:

$$\text{PG}_2 = 0.0594.$$

For Pier 3: $B_P = 11$ ft, $B_M = 35$ ft, $X = 262.5$ ft, $\sigma = \text{LOA} = 195$ ft

$$x_{\max} = \frac{262.5 + 0.5(11 + 35)}{195} = 1.4641,$$

$$x_{\min} = \frac{262.5 - 0.5(11 + 35)}{195} = 1.2282.$$

Find the area between 1.4641 and 1.2282 on the normal distribution curve having a mean at zero and a standard deviation of 1. Using statistical tables or Microsoft Excel with the excel function NORM.DIST, find the following:

$$\text{PG}_3 = 0.0424.$$

For Pier 4: $B_P = 11$ ft, $B_M = 35$ ft, $X = 597.5$ ft, $\sigma = \text{LOA} = 195$ ft

$$x_{\max} = \frac{597.5 + 0.5(11 + 35)}{195} = 3.1820,$$

$$x_{\min} = \frac{597.5 - 0.5(11+35)}{195} = 2.9462.$$

Find the area between 3.1820 and 2.9462 on the normal distribution curve having a mean at zero and a standard deviation of 1. Using statistical tables or Microsoft Excel with the excel function NORM.DIST, find the following:

$$\mathrm{PG}_4 = 0.0186.$$

Step 9. Determine the annual trips, *N*.

$$N = 10 \text{ trips/week} \times 52 \text{ weeks} = 520 \text{ trips}.$$

Step 10. Determine the distribution of annual frequency among the piers.

The waterway width is 1,345 ft. Six times the LOA is 6 × 195 = 1,170 ft. Three times the LOA is 3 × 195 = 585 ft. So, Pier 4 is the only pier outside three times the LOA. Strictly speaking, the LOA could seemingly be divided among Piers 1 through 3, with Pier 4 designed for the minimum collision load. However, since Pier 4 is not extremely far from the transit path centerline (597.5/195 = 3.06, only slightly larger than 3), we will divide the AF equally among all four piers for this example.

$$\text{Target AF} = 0.001/4 = 0.00025 \text{ at each pier}.$$

Step 11. Establish pier collision capacity values, H_P, to satisfy the required AF. Take the protection factor PF = 1.00, assuming no pier protection will be provided.

There is no single unique solution. One possible strategy is summarized in the calculations below.

At Pier 1:

$$\mathrm{AF} = N(\mathrm{PA})(\mathrm{PG})(\mathrm{PC})(\mathrm{PF}) = 520(0.000204)(0.0213)(\mathrm{PC})(1.00)$$

$$= 0.00226(\mathrm{PC}) = 0.00025$$

$$\mathrm{PC} = 0.111.$$

Back solve for the required resistance, *H*.

$$\mathrm{PC} = \begin{cases} 0.1 + 9\left(0.1 - \dfrac{H}{P}\right) & \text{if } 0.0 \le H/P < 0.10 \\ 0.111\left(1 - \dfrac{H}{P}\right) & \text{if } 0.10 \le H/P < 1.00 \\ 0.00 & \text{if } H/P \ge 1.00 \end{cases}.$$

Assume *H/P* is less than 0.10 and check:

$$0.111 = 0.1 + 9\left(0.1 - \frac{H}{P}\right) \rightarrow \frac{H}{P} = 0.0989 < 0.1, \text{ assumption was correct,}$$

$$H = 0.099 \times 2,248 = 222 \text{ kips.}$$

Similar calculations yield the following for the other piers:

Pier	PC	*H/P*	*H*, kips	AF
1	0.111	0.099	222	0.00025
2	0.040	0.644	1,771	0.00025
3	0.055	0.501	1,380	0.00025
4	0.124	0.097	211	0.00025
Total				0.00100

Given the consequences of bridge damage due to vessel collision on navigable waterways with high traffic volumes, it would seem prudent to adopt full pier protection (PF = 0) in many cases.

5 Blast Loading

5.1 PRELIMINARIES

Blast loading on structures often takes the form of impulsive, short-duration extreme pressure applied to the structure. Using principles of structural dynamics, there will be cases for which the structure does not "feel" the full dynamic pressure due to differences in the structure natural period and the duration of the loading.

Three very useful references, and the basis for much of the material presented here, for the analysis of structural loading due to blast effects are as follows:

- *Blast Protection of Buildings* (American Society of Civil Engineers, 2011),
- *Structures to Resist the Effects of Accidental Explosions* (United States Department of Defense, 2008), and
- *Design of Blast Resistant Structures* (Gilsanz et al., 2013).

The pressure created on a structural element from blast effects depends on the detonation velocity of the explosive, among other variables. Detonation velocities for modern explosives typically range from 22,000 to 28,000 ft/s.

Explosive effects may be produced from either unconfined or confined detonation. Subcategories exist within each of these two general categories of explosions.

Unconfined explosions may be as follows:

- free air burst, producing unreflected pressure loading on a nearby building; the explosive category least likely to occur,
- air burst, producing reflected pressure loading on a nearby building; an air burst explosion occurs at a distance above the ground that is two to three times the height of a two-story building; explosions significantly higher above the ground would be classified as free air burst explosions, and
- surface burst, producing reflected pressure on a nearby building; the initial shock is amplified due to ground reflection of the generated pressures.

Confined explosions may be as follows:

- fully vented, detonated in an open structure and producing internal shock loading,
- partially confined, detonated in a partial containment cell and producing internal shock loading, and
- fully confined, detonated in a full containment cell and producing internal shock loading.

DOI: 10.1201/9781003538363-5

TABLE 5.1
Heat of Detonation for Various Explosives

Explosive	H^d_{EXP} (ft-lb/lb)
TNT	1.97×10^6
Comp C-4	2.22×10^6
HMX	2.27×10^6
RDX	2.27×10^6
Boracitol	5.59×10^6
Tetryl	2.11×10^6

5.2 EXPLOSIVE MATERIALS

Explosive materials may be solid, liquid, or gas. Solid explosives are well understood in terms of blast pressures produced. Modern high explosives often are of the solid form, as other liquid and gaseous explosives often exhibit incomplete detonation, with a large amount of energy dissipated as heat (which may produce fires).

The basis material for much of the theory in blast pressure prediction is the explosive trinitrotoluene, or TNT. For other materials, an effective charge weight is determined using Equation 5.1.

$$W_E = \frac{H^d_{\text{EXP}}}{H^d_{\text{TNT}}} W_{\text{EXP}}. \tag{5.1}$$

The effective charge weight, W_E, for subsequent calculation of blast pressure depends on the actual weight of the explosive, W_{EXP}, and the respective values for heat of detonation for the explosive, H^d_{EXP}, and for TNT, H^d_{TNT}. Table 5.1 lists values for some explosives. For a more complete listing of this data, refer to UFC 3-340-02 (United States Department of Defense, 2008).

5.3 BLAST PRESSURE ANALYSIS – UNCONFINED DETONATION

The procedure for estimating blast pressures due to the detonation of unconfined explosives is limited in this text to the surface burst category, as this is the most likely category to occur. For treatment of free air burst and air burst unconfined detonations, refer to UFC 3-340-02 (United States Department of Defense, 2008).

With surface burst explosions, the initial incident wave of the explosion merges with a reflected wave (reflected by the ground surface). This merged wave is roughly hemispherical in shape. Compared to the free air burst and the air burst explosions, surface burst explosions are more severe for any given distance from the detonation.

Blast pressures generated form surface burst explosions are characterized by a positive phase and a negative phase. The negative phase is typically much smaller in pressure magnitude than the positive phase and may, at times, be disregarded in the design process.

An important parameter in blast pressure determination is the scaled distance, Z, given in Equation 5.2. W_{EXP} has been previously defined as the equivalent explosive weight. The minimum distance from the detonation center to the structural element in question (e.g., a wall facing the explosion) is denoted as R (ft).

$$Z = \frac{R}{(W_{EXP})^{1/3}}. \tag{5.2}$$

Figure 5.1 (from UFC 3-340-02 Figure 2.15) may be used to establish the parameters for a surface burst explosive detonation.

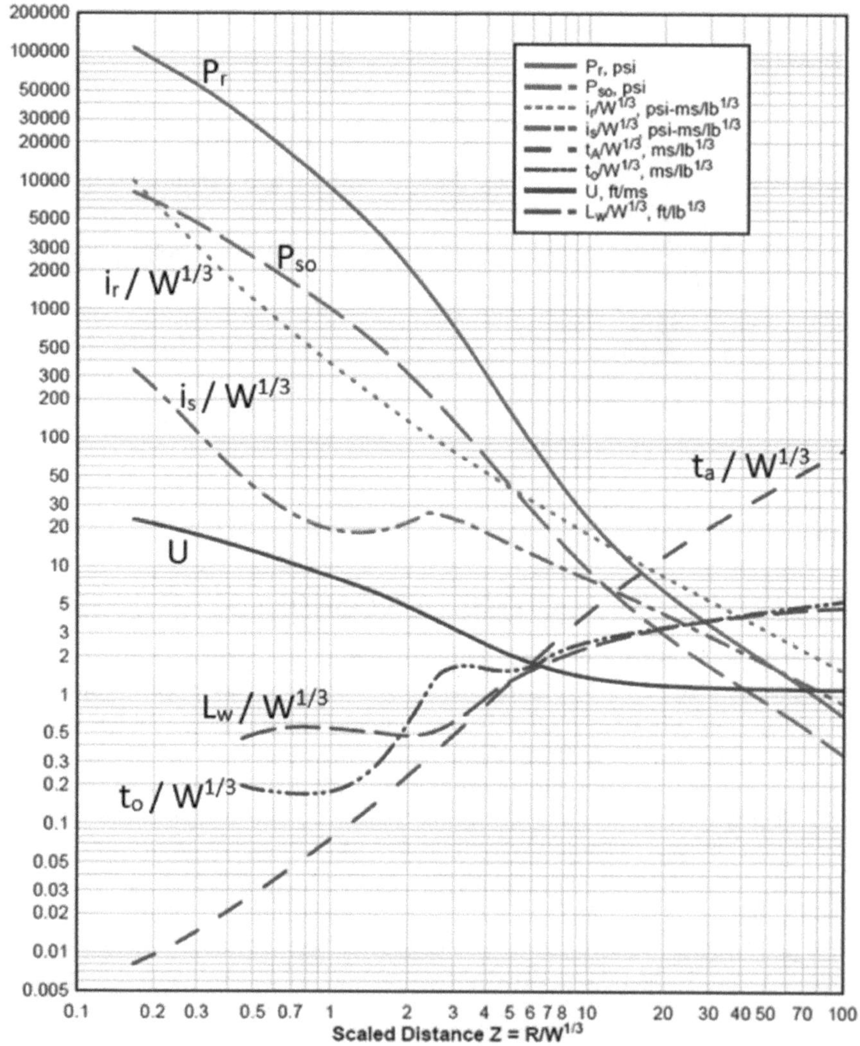

FIGURE 5.1 Positive phase surface burst pressure and impulse parameters.

The parameters determined from Figure 5.1 are defined as follows:

- P_r = peak reflected pressure, psi,
- P_{so} = peak incident pressure, psi,
- i_r = reflected impulse, psi-ms,
- i_s = incident impulse, psi-ms,
- t_A = time elapsed from detonation to pressure wave incidence, ms,
- t_o = positive phase duration, ms,
- U = shock velocity, ft/ms, and
- L_w = positive wave length; the length from the detonation which experiences positive pressure at a particular instant of time, ft.

With regard to pressure loading, an impulse is the area beneath the pressure–time curve. Figure 5.2 is a simplified pressure–time curve frequently adopted in design for the positive phase only (American Society of Civil Engineers, 2011). Some of the following additional parameters are required for full definition of the loading.

- t_c = average clearing time to relieve reflected pressure, seconds,
- t_{of} = duration of equivalent triangular incident loading, seconds,
- t_{rf} = fictitious duration of equivalent triangular reflected loading, seconds,
- S = clearing distance, ft; equal to the lesser of the structure height, H, or one-half of the structure width, W_S, perpendicular to the direction of loading,
- G = the larger of the structure height, H, or one-half of the structure width, W_S, perpendicular to the direction of loading,

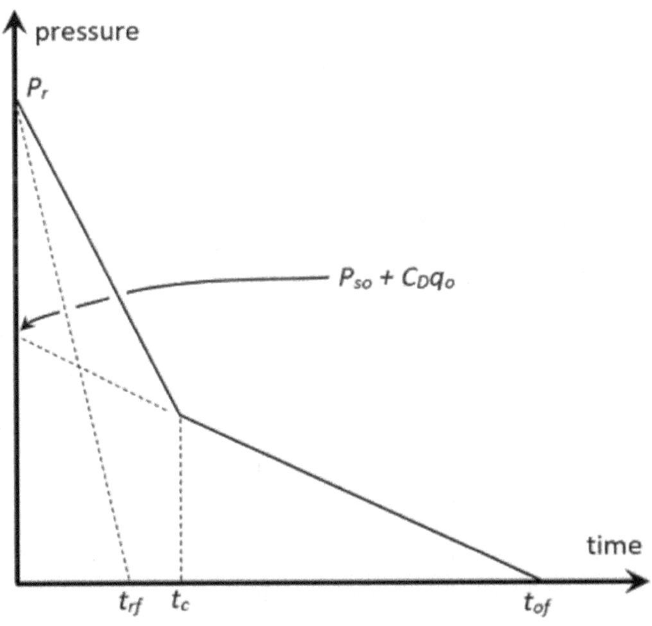

FIGURE 5.2 Positive phase pressure–time curve for surface burst explosions.

- C_D = drag coefficient; equal to 1.0 for the front wall exposed to an explosion, and
- C_r = velocity of sound in the reflected region, ft/ms.

The velocity of sound depends on the peak incident over-pressure and may be obtained from Figure 5.3 (Figure 2-192 in UFC 3-340-02). The average clearing time, t_c, is given in Equation 5.3.

$$t_c = \frac{4S}{C_r(1+S/G)}. \tag{5.3}$$

The procedure outlined in UFC 3-340-02 includes a recommendation that the effective charge weight be amplified by 20% for a safety factor.

Since these equivalent loadings are triangular, the durations are easily determined from the impulse and pressure values derived from Figure 5.2.

$$t_{\mathrm{of}} = \frac{2i_s}{P_{\mathrm{so}}}, \tag{5.4}$$

$$t_{\mathrm{rf}} = \frac{2i_r}{P_r}. \tag{5.5}$$

The material presented here has focused on surface burst explosions and their effects on directly loaded structural elements, such as the front wall of a building. For a detailed treatment of blast loading on all surfaces of a structure, confined explosions, fragmentation produced from explosions, and other related topics, the reader is referred to UFC 3-340-02.

5.4 DESIGN CRITERIA FOR STRUCTURES

Design criteria for structures subject to explosive loading may be found in multiple documents and standards.

The Commentary to ASCE 59-11 (American Society of Civil Engineers, 2011) provides recommended limits for components based on displacement ductility and support rotation estimates. UFC 3-340-02 contains extensive full chapters on requirements for reinforced concrete, structural steel, masonry, precast concrete, connections, pre-engineered buildings, blast resistant windows, and underground structures.

5.5 DYNAMICS OF STRUCTURES SUBJECT TO IMPULSIVE LOADS

When a structure or component may be idealized as a single-degree-of freedom (SDOF) system, there are at least three methods that can be used to solve for the response of a linear system to impulsive loading.

- numerical, step-by-step solution of the equation of motion,
- closed-form solutions for undamped systems, and
- approximate, simplified impulse load analysis when the load duration is short relative to the natural period of the system.

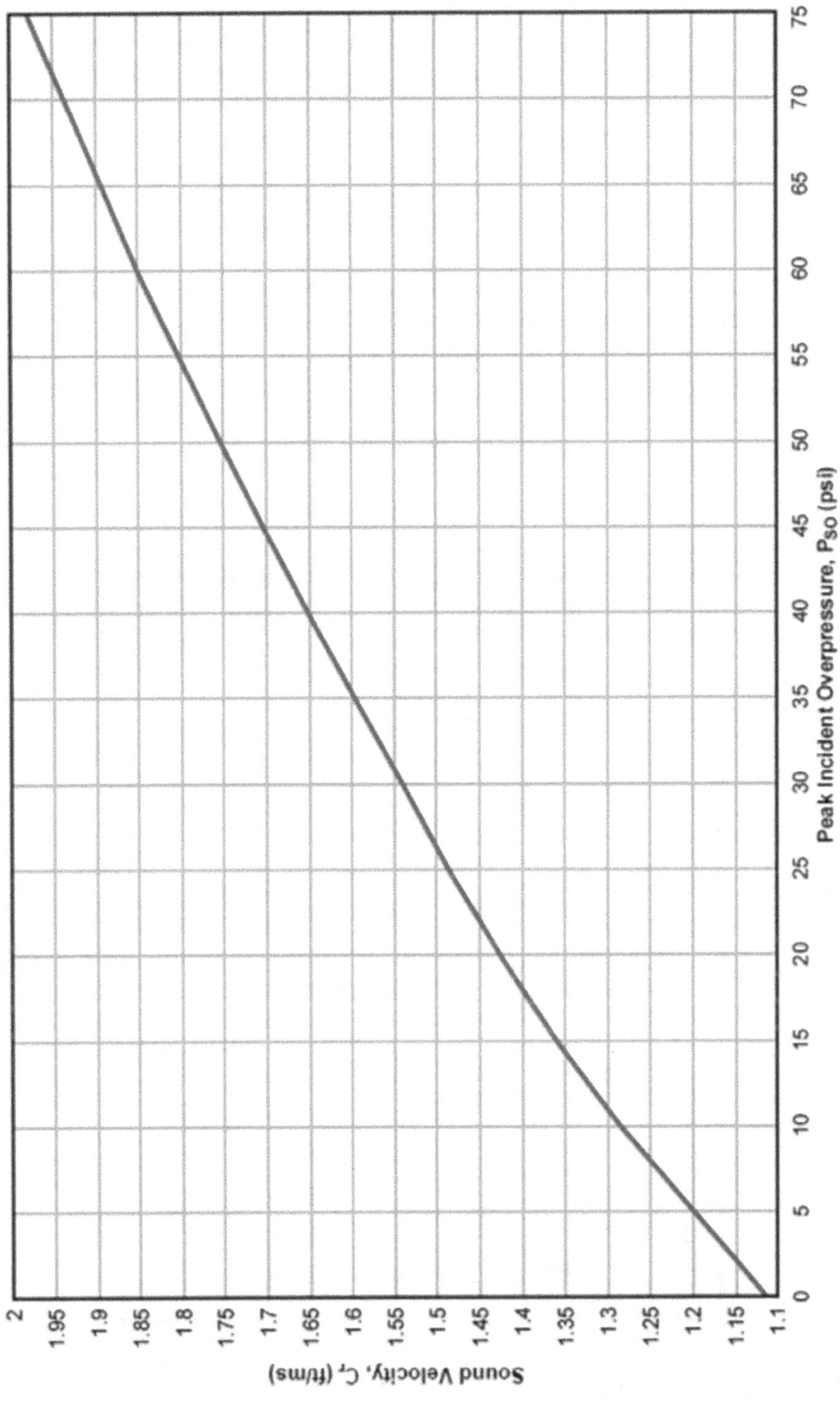

FIGURE 5.3 Velocity of sound vs. peak incident over-pressure.

These three methods are generally linear in nature and do not account for inelastic behavior, which is typically required for elements subject to blast loading. Nonetheless, the response of an elastic system is useful in making sense of results from inelastic analyses.

The advantages in numerical solution of the equation of motion over the other two methods are (1) the potential for including damping, and (2) the potential for including nonlinear stiffness. Closed-form solutions presented here and the approximate method presented here are for undamped systems. For linear systems (constant stiffness, k), Chopra (2016) presents a set of recurrence formulas that involve eight constants. These constants need to be computed only once, and the step-by-step solution for both displacement and velocity is determined as shown in Equations 5.6 and 5.7. For the constants (A, B, C, D, A', B', C', and D'), the reader is referred to the excellent book by Chopra.

$$u_{i+1} = Au_i + B\dot{u}_i + Cp_i + Dp_{i+1}, \tag{5.6}$$

$$\dot{u}_{i+1} = A'u_i + B'\dot{u}_i + C'p_i + D'p_{i+1}. \tag{5.7}$$

The displacement is u and the force at a given instant of time is p.

Closed-form solutions for undamped systems subjected to impulsive loads of various shapes are available in most structural dynamics textbooks. For the triangular pulse shown in Figure 5.4, one shape which may be useful for the analysis of blast loading, Equations 5.8 through 5.12 provide a solution for the undamped

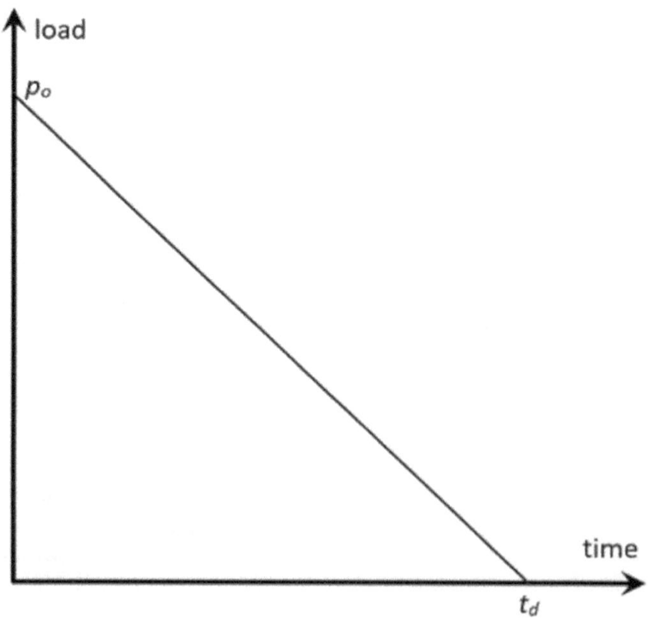

FIGURE 5.4 Unsymmetric triangular impulsive loading.

displacement response, $u(t)$, of a SDOF system as a function of time. Solutions for undamped systems subjected to short-duration loads are often excellent estimates of the actual damped response.

$$u = \frac{p_o}{k}\left[\frac{\sin \omega_n t}{\omega_n t_d} - \cos \omega_n t - \frac{t}{t_d} + 1\right], \quad t \le t_d, \tag{5.8}$$

$$u = \frac{C_1}{\omega_n}\sin(\omega_n t - \omega_n t_d) + C_2 \cos(\omega_n t - \omega_n t_d), \quad t \ge t_d, \tag{5.9}$$

$$C_1 = \frac{p_o \omega_n}{k}\left[\frac{\cos(\omega_n t_d)}{\omega_n t_d} + \sin(\omega_n t_d) - \frac{1}{\omega_n t_d}\right], \tag{5.10}$$

$$C_2 = \frac{p_o}{k}\left[\frac{\sin(\omega_n t_d)}{\omega_n t_d} - \cos(\omega_n t_d)\right], \tag{5.11}$$

$$\omega_n = \sqrt{\frac{k}{m}}, \tag{5.12}$$

where k is the SDOF system stiffness, m the SDOF system mass, ω_n the natural frequency of the undamped system, radians per second, p_o the load value at time zero, and t_d the load duration.

When the loading shape is triangular including a rise time, as shown in Figure 5.5, the closed-form solution for the undamped SDOF system is given in Equations 5.13 through 5.15.

$$u = \frac{2p_o}{k}\left[\frac{t}{t_d} - \frac{\sin(\omega_n t)}{\omega_n t_d}\right], \quad t \le {}^{t_d}\!/_2, \tag{5.13}$$

$$u = \frac{2p_o}{k}\left[1 - \frac{t}{t_d} + \frac{1}{\omega_n t_d}\left(2\sin\left(\omega_n t - \frac{\omega_n t_d}{2}\right) - \sin(\omega_n t)\right)\right], \quad {}^{t_d}\!/_2 \le t \ge t_d, \tag{5.14}$$

$$u = \frac{2p_o}{k}\left[\frac{1}{\omega_n t_d}\left(2\sin\left(\omega_n t - \frac{\omega_n t_d}{2}\right) - \sin(\omega_n t - \omega_n t_d) - \sin(\omega_n t)\right)\right], \quad t \ge t_d. \tag{5.15}$$

For a loading defined by a half-sine pulse, as shown in Figure 5.6, the closed-form solution for an undamped system is given in Equations 5.16 and 5.17 for the case in which t_d does not equal $T_n/2$. The forcing frequency ratio for the half-sine pulse is given in Equation 5.18. For the special case in which $t_d = T_n/2$ ($\beta = 1$), Equations 5.19 and 5.20 provide the theoretical undamped solution.

$$u = \frac{p_o}{k}\left(\frac{1}{1-\beta^2}\right)\left(\sin\frac{\pi t}{t_d} - \beta \sin\frac{2\pi t}{T_n}\right) \quad \text{for } t \le t_d \quad \text{and} \quad \beta \ne 1, \tag{5.16}$$

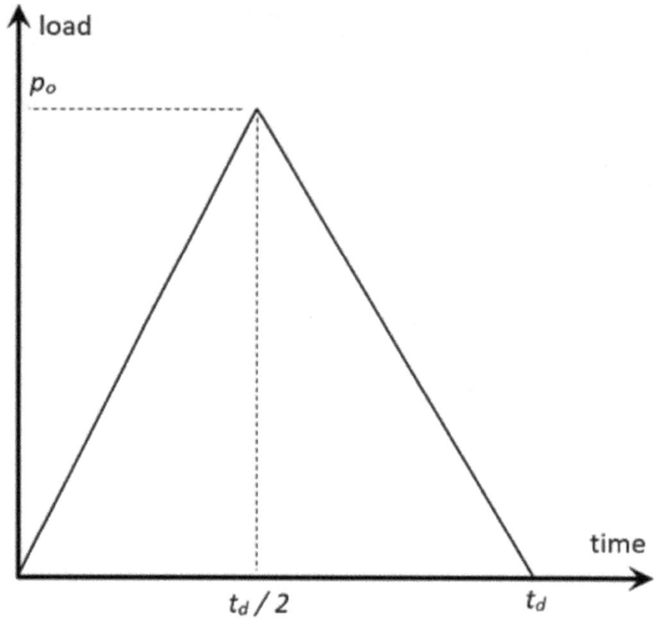

FIGURE 5.5 Triangular pulse with rise time.

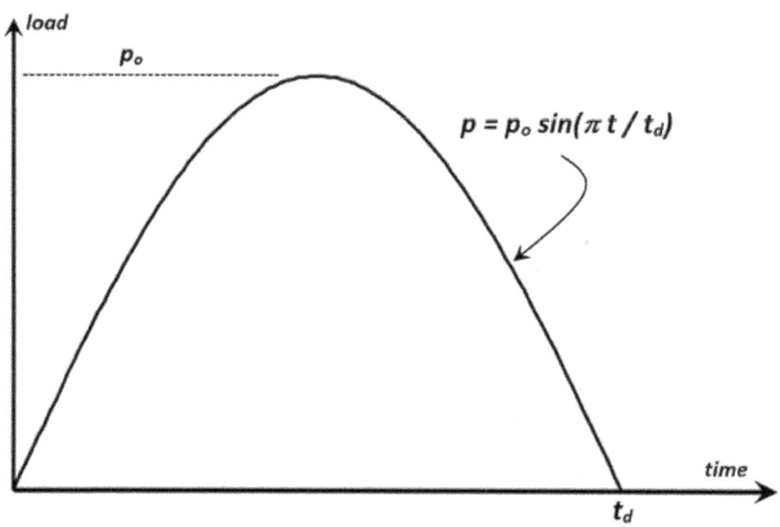

FIGURE 5.6 Half-sine pulse.

$$u = \frac{p_o}{k}\left(\frac{2\beta\cos\dfrac{\pi}{2\beta}}{\beta^2 - 1}\right)\sin\left[2\pi\left(\frac{t}{T_n} - \frac{t_d}{2T_n}\right)\right] \quad \text{for } t \geq t_d \quad \text{and} \quad \beta \neq 1, \tag{5.17}$$

$$\beta = \frac{T_n}{2t_d}, \tag{5.18}$$

$$u = \frac{p_o}{2k}\left(\sin\frac{2\pi t}{T_n} - \frac{2\pi t}{T_n}\cos\frac{2\pi t}{T_n}\right) \quad \text{for } t \leq t_d \quad \text{and} \quad \beta = 1, \tag{5.19}$$

$$u = \frac{\pi p_o}{2k}\cos\left(\frac{2\pi t}{T_n} - \pi\right) \quad \text{for } t \geq t_d \quad \text{and} \quad \beta = 1. \tag{5.20}$$

When the load duration, t_d, is short relative to the natural period of the structure, $T_n = 2\pi/\omega_n$, the shape of the impulsive load becomes less critical and an approximate solution for the maximum displacement of an undamped SDOF system is given in Equation 5.21. The impulse, given by the integral in Equation 5.22, is simply the area under the load–time curve.

$$u \cong \frac{i}{m\omega_n}\sin\left[\omega_n(t - t_d)\right], \tag{5.21}$$

$$i = \int_0^{t_d} p\,dt. \tag{5.22}$$

This approximate solution neglects damping and gives reasonable approximations to the true value only when t_d/T_n is about 0.25 or less.

5.6 ILLUSTRATIVE EXAMPLES

Suppose that a detonation of an explosive with equivalent charge weight, $W_E = 750$ lbs, occurs 40 ft away from a building element idealized as a SDOF with a weight of 54 kips and an initial stiffness of 10 kips/inch. The post-yield stiffness ratio of the equivalent SDOF system is zero (elastic-perfectly-plastic). The structural element on the front of the building is directly loaded by the blast wave and has a tributary area equal to 350 ft^2. The loading is to be idealized using the unsymmetric triangular pulse in Figure 5.4. Determine the required yield force of the element if the displacement ductility is to be limited to no more than 5.

Determine the scaled distance, Z. Use a 20% increase in the charge weight as a safety factor.

$$W_{\text{EXP}}^{1/3} = (1.2 \times 750)^{1/3} = 9.65 \ \text{lb}^{1/3},$$

$$Z = \frac{R}{(W_{\text{EXP}})^{1/3}} = \frac{40}{(1.2 \times 750)^{1/3}} = 4.14 \text{ ft/ lb}^{1/3}.$$

From Figure 5.1 determine the following:

- P_r = peak reflected pressure = 200 psi,
- P_{so} = peak incident pressure = 45 psi,
- i_r = reflected impulse = 33 × 9.65 = 319 psi-ms,
- i_s = incident impulse = 13 × 9.65 = 136 psi-ms, and
- t_A = time elapsed from detonation to pressure wave incidence = 1 × 9.65 = 9.65 ms.

The fictitious duration, t_{rf}, is given in Equation 5.5.

$$t_{rf} = \frac{2i_r}{P_r} = \frac{2(319)}{200} = 3.19 \text{ ms} = 0.00319 \text{ seconds}.$$

The natural period of the structure may be computed from the given mass and stiffness parameters.

$$T_n = 2\pi\sqrt{\frac{W}{gk}} = 2\pi\sqrt{\frac{54}{386 \times 10}} = 0.743 \text{ seconds}.$$

The initial load is the pressure multiplied by the tributary area of the element.

$$p_0 = 200 \times (350 \times 144) \div 1{,}000 = 10{,}080 \text{ kips}.$$

Blast analysis often must incorporate nonlinear effects and damping in the solution to be accurate. Multi-degree-of-freedom (MDOF) software, available at the *National Hazards Engineering Research Infrastructure* DesignSafe repository, is instructive for such purposes.

https://simcenter.designsafe-ci.org/products/learning-tools/multiple-degrees-of-freedom/

MDOF has capabilities for MDOF systems, but may also be used to analyze simple, single-degree-of-freedom systems incorporating both damping and yielding.

As a second example, focusing on dynamic structural response to impulsive blast loading, consider a single-degree-of-freedom system weighing 260 kips, with an initial stiffness equal to 130 kips/inch. The system is subjected to the blast load shown in Figure 5.7. Solve for the maximum displacement of the system for the following five cases:

1. using the approximate method based on the impulse (damping ignored),
2. using the undamped, closed-form solution for an equivalent half-sine pulse (with the same impulse and duration as the actual load),
3. using MDOF software for an undamped system with infinite yield force,
4. using MDOF software for a system with damping equal to 31% of critical damping and an infinite yield force, and
5. using MDOF software for a system with 31% damping, a post-yield stiffness equal to zero (elastic-perfectly-plastic), and a yield force equal to 45 kips.

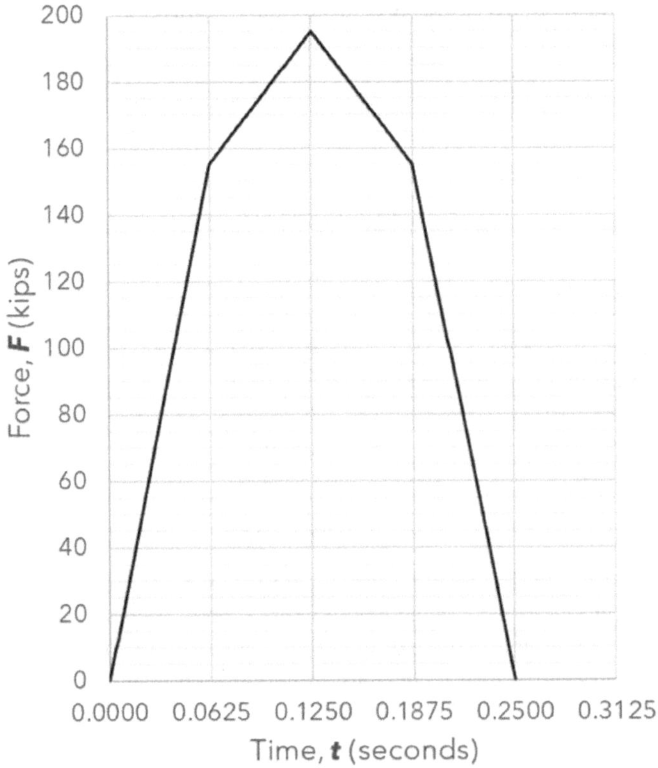

FIGURE 5.7 Blast force example.

5.6.1 Case 1: Approximate Solution

$$i = \frac{1}{2}(155)(0.0625)(2) + 175(0.0625)(2) = 31.5625 \text{ k-sec},$$

$$u_{\max} \cong \frac{31.5625}{13.89(260/386)} = 3.37 \text{ inches.}$$

5.6.2 Case 2: Equivalent Half-Sine Pulse

An equivalent half-sine pulse has the same area under the load–time curve as does the actual loading and has the same duration as the actual loading. It may be shown that the amplitude of an equivalent half-sine pulse is $p_o = (i\pi)/(2t_d)$.

$$p_o = \frac{31.5625\pi}{2(0.25)} = 198.31 \text{ kips.}$$

Note that the natural period of the system, T_n, is as follows:

$$T_n = 2\pi\sqrt{\frac{260}{130(386)}} = 0.452 \text{ seconds}.$$

Hence, the ratio:

$$t_d / T_n = 0.25 / 0.452 = 0.553.$$

Since t_d/T_n is significantly greater than 0.25, the approximate solution would generally not be reliable in this particular problem. Applying the closed-form solution produces a maximum displacement, u_{max}, for the half-sine pulse. It should be no surprise that the two solution completed thus far do not agree.

$$u_{\max} = 2.51 \text{ inches}.$$

5.6.3 Case 3: MDOF with No Damping and No Yielding

A load curve corresponding to the given loading may be generated for MDOF, and an analysis for the SDOF system can be completed, producing another estimated displacement for the undamped, unyielding system. The solutions for the half-sine pulse (2.51 inches) and MDOF (2.48 inches) are in good agreement.

$$u_{\max} = 2.48 \text{ inches}.$$

5.6.4 Case 4: MDOF with 31% Damping and No Yielding

The given damping equal to 31% of critical is very high and indicative of an added damping system. Typical damping values for conventional structural systems are in the range of 1%–7% of critical damping. Nonetheless, the effect of damping incorporated into the MDOF model produces a damped displacement estimate.

$$u_{\max} = 1.66 \text{ inches}.$$

5.6.5 Case 5: MDOF with 31% Damping and $F_Y = 45$ kips

Since systems are often designed to experience damage in the form of yielding when subjected to blast forces, it will frequently be necessary to rely on procedures that incorporate both damping and yielding in the solution. MDOF was used to estimate the maximum displacement, including both effects. Figure 5.8 shows the displacement history of the system out to 20 seconds. The figure illustrates another important point – systems subjected to blast loading may have large permanent residual displacements, u_{res}, after the loading has ended.

$$u_{\max} = 3.19 \text{ inches},$$

$$u_{\text{res}} = 2.85 \text{ inches}.$$

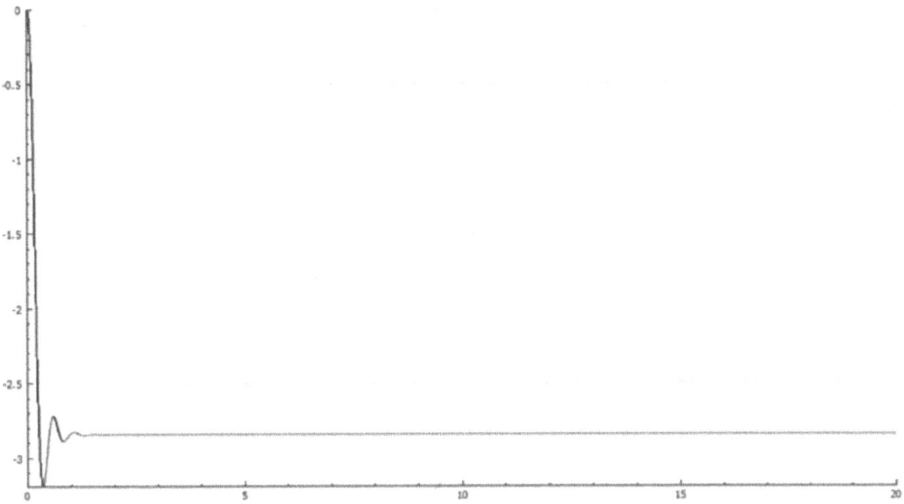

FIGURE 5.8 Example of displacement response history.

The yield displacement, u_y, and ductility demand, μ, on the system subjected to the given loading are important parameters in assessing the adequacy of a design incorporating nonlinear response.

$$u_y = \frac{45 \text{ kips}}{130 \text{ k/inch}} = 0.346 \text{ inches},$$

$$\mu = \frac{3.19}{0.346} = 9.22.$$

Various approximate solutions can be very simple to obtain and provide a good starting point for estimating response to blast loading. Final design typically needs to incorporate explicit damping and nonlinearities to obtain an accurate solution.

5.7 BLAST-GENERATED MISSILE EFFECTS

In addition to the pressure loading resulting from a blast, the effects from missile penetration and fragmentation have been the subject of extreme loadings on structures.

Several studies have included testing to assess the penetration of a missile into soil and rock. Young's equations, Equations 5.23 and 5.24 (Bulson, 1997), represent estimates of penetration depth, p, of a falling bomb onto various materials. The scatter in such estimates is extremely wide.

$$p = \left(0.53SN\sqrt{\frac{W_P}{A}}\right)\ln\left(1+\frac{2V^2}{10^5}\right) \quad \text{for } V < 200 \text{ ft/s}, \tag{5.23}$$

$$p = \left(0.0031SN\sqrt{\frac{W_P}{A}}\right)(V-100) \quad \text{for } V \geq 200 \text{ ft/s}, \tag{5.24}$$

TABLE 5.2
Soil Constant for Various Material Descriptions

Material Description	S
Rock	1.07
Dense, dry silty sand	2.50
Silty clay	5.20
Loose, moist sand	7.00
Moist clay	10.5
Wet silty clay	40.0
Soft wet clay	50.0

TABLE 5.3
Nose Coefficient for Various Projectile Shapes

Projectile Nose Shape	N
Flat nose	0.56
Tangent ogive 2.2 CRH	0.82
Tangent ogive 6 CRH	1.00
Cone, $L/D=3$	1.32

where p is the penetration in ft; W_P is the projectile weight, lbs; A is the cross-sectional area of projectile, inch^2; V is the striking velocity, ft/s; S is the soil constant (see Table 5.2); and N is the nose coefficient (see Table 5.3).

A range of validity has been suggested in Bulson (1997) for Young's equations:

- $W_P=2$–3,750 lbs,
- Projectile diameter = 1–30 inches,
- $W_P/A=0.1$–38 psi,
- $V=100$–2,370 ft/s.

For penetration of a striking missile into concrete, Bulson (1997) provides Equation 5.25.

$$p_c=\left(\frac{870}{\sigma_c^{0.5}}\right)\left(\frac{W_P}{d^2}\right)\left(\frac{d}{c}\right)^{0.1}\left(\frac{V}{1{,}750}\right)^n, \tag{5.25}$$

$$n=\frac{10.7}{\sigma_c^{0.25}}. \tag{5.26}$$

where p_c is the penetration into concrete, inches; W_P is the projectile weight, lbs; d is the projectile diameter, inches; σ_c is the crushing strength of concrete, psi V is the striking velocity, ft/s; and c is the maximum aggregate size, inches.

Bulson (1997) presents several other expressions for missile penetration and discusses fragmentation as well. For a more detailed study of this subject, the reader is directed to Bulson (1997).

6 Ice Loading

The AASHTO LRFD Bridge Design Specifications (AASHTO, 2020), in Section 3.9, include provisions for determining ice loads on bridge piers. The AASHTO provisions are applicable to freshwater ice on rivers and lakes. Ice loads in seawater must be determined "by suitable specialists using site-specific information".

Forces on bridge piers from ice loads are both dynamic and static in nature. The magnitude of dynamic ice force depends on the strength of the flowing ice. AASHTO provides the following ice crushing strength data to be used unless more site-specific information is available.

- 8.0 ksf, where breakup occurs at melting temperatures and the ice structure is substantially disintegrated. This ice strength is appropriate when ice forces are expected to be minimal but an allowance is to be reasonably required for ice loading.
- 16.0 ksf, where breakup occurs at melting temperatures and the ice structure is somewhat disintegrated.
- 24.0 ksf, where breakup or major ice movement occurs at melting temperatures, but the ice moves in large pieces and is internally sound.
- 32.0 ksf, where breakup or major ice movement occurs when the ice temperature, averaged over its depth, is measurably below the melting point. While this is viewed in many circles as a reasonable upper limit on ice strength, values as high as 57.6 ksf have been used in some cases in Alaska.

These ice strength values are not necessarily representative of actual values form laboratory testing (which are often considerably higher) but are somewhat notional in nature and intended to be used in the subsequent equations.

The pertinent parameters used in calculating ice loads on bridge piers are as follows:

> t is the ice thickness, ft; w is the pier width at the level of ice action, ft; p is the effective ice crushing strength, ksf; α is the inclination of the nose to vertical, degrees; F_c is the horizontal ice force caused by ice floes that fail by crushing over the full width of the pier, kips; F_b is the horizontal ice force caused by ice floes that fail by flexure as they ride up the inclined pier nose, kips; C_a is the coefficient accounting for the effect of the pier width/ice thickness ratio where the floe fails by crushing; C_n is the coefficient accounting for the inclination of the pier nose with respect to vertical.

For cases in which the angle of inclination, α, is less than or equal to 15°, ice failure by flexure is not considered to be a possible ice failure mode, and F is taken as F_c.

DOI: 10.1201/9781003538363-6

If w/t is greater than 6, then the ice force $F = F_c$. Otherwise, the ice force F is the lesser of F_c and F_b.

$$F_c = C_a ptw, \tag{6.1}$$

$$F_b = C_n pt^2, \tag{6.2}$$

$$C_a = \sqrt{1 + 5t / w}, \tag{6.3}$$

$$C_n = \frac{0.50}{\tan(\alpha - 15)}. \tag{6.4}$$

The largest source of uncertainty in estimating ice forces on bridge piers is the ice thickness, t, which should be based on extreme – not average or median – values from site-specific data when available. When such data are unavailable, the AASHTO Commentary suggests an empirical method given in Equations 6.5 and 6.6. The parameter, k_t, is adopted here in lieu of the symbol used in the AASHTO Commentary to avoid confusion with the pier nose inclination, α.

The so-called freezing index, S_f, is summed from the date of freeze-up to the date of interest and has units of degree-days. The mean daily temperature, T, is to be expressed in degrees Fahrenheit.

$$t = 0.083(k_t)\sqrt{S_f}, \tag{6.5}$$

$$S_f = \sum (32 - T). \tag{6.6}$$

Suggested values of k_t from the AASHTO Commentary are as follows:

- windy lakes without snow, $k_t = 0.8$,
- average lake with snow, $k_t = 0.5$–0.7,
- average river with snow, $k_t = 0.4$–0.5, and
- sheltered small river with snow, $k_t = 0.2$–0.4.

It is unrealistic to expect ice forces form ice floes to align perfectly with the pier axis. Therefore, AASHTO requires that, even when the flow theoretically aligns with the axis of the pier, a transverse ice load acts in unison with the longitudinal ice load. Two cases are to be considered when the theoretical force aligns with the pier:

- Case 1: A longitudinal force, F, calculated form Equations 6.1–6.4 combined with a transverse force equal to $0.15F$, and
- Case 2: A longitudinal force equal to $0.5F$, where F is calculated from Equations 6.1–6.4, combined with a transverse force given in Equation 6.7.

$$F_t = \frac{F}{2\tan(\beta / 2 + \theta_f)}. \tag{6.7}$$

β is the pier nose angle in a horizontal plane, taken to be 100° for a round-nosed pier. θ_f is the friction angle (degrees) between the pier and the ice. No guidance is provided in AASHTO regarding the friction angle.

When the pier is skewed relative to the direction of flow, the ice force is to be computed using Equations 6.1–6.4, with the projected width of the pier taken as w, and the resulting force broken into transverse and longitudinal components. In such cases, the transverse force may not be taken less than 20% of the longitudinal force.

Static loads from ice on piers include those from (1) static pressure due to thermal movements of ice sheets, (2) hanging dams or jams of ice, and (3) static uplift or vertical load resulting from adhering ice in waters of fluctuating levels. For cases (1) and (2), refer to Section 3.9.5 of the AASHTO LRFD Bridge Design Specifications and the literature referenced there. For case (3), the vertical force on a bridge pier from ice adhesion in fluctuating waters is given in Equation 6.8 for a circular pier and in Equation 6.9 for an oblong pier.

$$F_v = 80t^2\left(0.35 + \frac{0.03R}{t^{0.75}}\right), \tag{6.8}$$

$$F_v = 0.20t^{1.25}L + 80t^2\left(0.35 + \frac{0.03R}{t^{0.75}}\right). \tag{6.9}$$

where t is the ice thickness, ft; R is the radius of a circular pier, ft; or radius of half circles at ends of an oblong pier, ft; or radius of a circle that circumscribes each end of an oblong pier of which the ends are not circular in plan at water level, ft; L is the perimeter of pier, excluding half circles at ends of oblong pier, ft.

Collins (2013) adopted a probabilistic approach based on measured data over 11 years to determine ice loads experienced by bridge piers of the Confederate Bridge in Canada. The Confederate Bridge is 11 km long and consists of 44 piers subject to heavy ice loading. Measurements were based on pier response to impacting ice using tilt meters. The maximum measured ice load over the 11-year time period was 6.7 Mega Newtons (MN; 1,506 kips). The estimated pier capacity in the same study was 24 MN (5,395 kips). In a separate report (Barrette et al., 2017), ice thickness reported at the Confederate Bridge was as high as 1.2 m (47 inches) during the same span of time.

With regard to ice flow forces in rivers, the literature includes a review of several different methods for such calculations (Rodtang et al., 2023).

7 Earthquake Ground Motion

7.1 PRELIMINARIES

The goal of this book is to provide direction in the loading of structures subjected to extreme events, with ground shaking from earthquakes the last to be considered. For this topic, the focus is on ground shaking effects. Other effects, such as liquefaction and lateral spreading, are not considered here. Loading of structures due to ground shaking from earthquakes is typically defined as either (1) a design response spectrum or (2) a suite of carefully selected and modified ground motion records. In either case, the design response spectrum is required since it provides one criterion for the selection and modification of ground motion records.

Generally, the assessment of ground shaking on structures is accomplished using either (1) static, equivalent force analysis, (2) linear dynamic response spectrum analysis, (3) linear dynamic response history analysis, or (4) nonlinear dynamic response history analysis. In the first two cases, only the design response spectrum is needed to define the earthquake loading. In the latter two cases, both a design response spectrum and a suite of ground motion records are required to define the earthquake loading.

The majority of this chapter is devoted to methods used to develop design response spectra and ground motion suites in accordance with various codes and standards. Nonetheless, a basic understanding of the nature of earthquake ground motion is a prerequisite to full appreciation of loading specifications for structural analysis and design.

A brief outline will assist in navigating this lengthy chapter.

Section 7.2 provides a discussion of ground motion characterization. Included are example problems covering seismic site classification and response spectrum generation using alternative site factors for a hypothetical project in the Mississippi Embayment of the New Madrid seismic zone.

Section 7.3 includes features of ground motion adjustment and filtering to remove noise recorded along with the ground shaking. An example of unfiltered and filtered characteristics of ground motion recording is included.

Section 7.4 provides a summary of several ground motion databases available to scientists and engineers to assist in developing ground motion suites.

Section 7.5 highlights several ground motion parameters used in assessing the intensity of a given ground motion recording.

Section 7.6 covers seismic loading in accordance with ASCE 7 (ASCE, 2022). Included are site classification procedures from both ASCE 7-16 and ASCE 7-22. ASCE 7 criteria for ground motion selection and modification are included as well. An example for ground motion modification by spectral matching for linear response history analysis is presented.

DOI: 10.1201/9781003538363-7

Section 7.7 provides a more in-depth coverage of ground motion response spectra, both elastic and inelastic. The important issue of damping in analytical techniques is touched upon. The substitute structure method is introduced as a valuable tool in estimating nonlinear behavior using an equivalent linear analysis. An example problem is presented, which compares results from nonlinear response history analysis to those obtained using the substitute – structure method. This section also includes a summary of online tools available for assistance in defining design response spectra.

Section 7.8 covers the use of artificial and synthetic ground motion records. There may be times when no recordings from actual earthquakes can be identified as appropriate for a particular project and site. In such cases, a viable option may be to incorporate artificial and synthetic records.

Section 7.9 presents a brief discussion on statistical features of design earthquake ground motion.

Section 7.10 summarizes various ground motion models which have been developed by researchers over the past decade. These models are often used in establishing seismic hazard analyses to define design response spectra. A simple example using three scenarios for a given site and ground motion model is included.

Section 7.11 covers earthquake loading as defined by AASHTO (AASHTO, 2011) for bridges. An example problem including definition of design response spectra as well as ground motion selection and modification is included.

Section 7.12 covers earthquake loading as defined by ASCE 43 (ASCE, 2019) for nuclear safety-related structures. An example problem that includes variability in subsurface conditions incorporated in the development of design response spectra is covered in detail. This example includes methods used to convert design response spectra from geometric mean-based to maximum-direction-based as well.

Section 7.13 is a brief discussion of seismic loading as defined by ASCE 41 (ASCE, 2017) for the evaluation and retrofit of structures.

Section 7.14 presents coverage of structural criteria for earthquake loading. This includes the concept of risk category as found in ASCE 7, equivalent lateral force procedures, modal analysis and superposition, linear response history analysis, and nonlinear response history analysis. This section also delved more deeply into the procedures required in ground motion selection and modification. The methods used in ground motion modification – amplitude scaling, spectral matching in the frequency domain, and spectral matching in the time domain – are covered in this section more robustly than in previous sections. Another ground motion suite development example is included in this section. A second example highlights methods used to establish damping coefficients for Rayleigh damping and period range of interest for ground motion selection and modification for a particular structure. A third example presents the seismic analysis of a simple structure using four different methods and compares those results. Final topics for this section include brief discussions on lateral force-resisting systems for both buildings and for bridges. Basic calculations for a concentrically braced frame and for an eccentrically braced frame are presented.

7.2 GROUND MOTION CHARACTERIZATION

Many parameters affect the ground motion experienced at a given site from an earthquake. And the scatter in the ground motion experienced is significant. A magnitude seven earthquake could produce very different ground shaking at two sites having similar subsurface properties and located similar distances from the fault rupture. Some of the important parameters affecting ground shaking characteristics include magnitude, tectonic setting, source-to-site distance, and subsurface conditions. These four parameters are not the only influences, but the discussion here will be limited to these.

7.2.1 Magnitude

The specification of earthquake severity has taken on various definitions through time. These include the Richter scale, body wave magnitude, surface wave magnitude, and moment magnitude.

In the 1930s, Charles Richter developed an earthquake intensity scale based on the wave amplitude recorded on a seismometer combined with the distance from the earthquake source to the seismometer site. The Richter scale has also been called the local magnitude scale and given the symbol M_L.

Multiple scales based on body wave or surface wave measurements have been devised. The most frequently used scale in modern times has become the moment magnitude scale, which relates the energy released through faulting surfaces to a magnitude value, denoted M_W. Given the nature of the moment magnitude scale, it may be said that an increase of 1.0 in magnitude represents an increase by a factor of $10^{1.5} \approx 32$ in energy release. Equation 7.1 is the expression for moment magnitude as a function of seismic moment, defined as the rock strength times the rupture area times the slip (the distance the fault moved). In Equation 7.1, the seismic moment, M_o, needs to be expressed in units of dyne-cm. One pound is equal to 444,822.16 dynes:

$$M_W = \frac{2}{3} \cdot \log_{10} M_o - 10.7 \tag{7.1}$$

7.2.2 Tectonic Setting

Tectonic regimes refer to the type of faulting which occurs within a given system. Regimes may be generally characterized as shallow crustal and subduction earthquakes. Within the shallow crustal category are included strike-slip, reverse, normal, and oblique faults. Subduction regions are characterized by one tectonic plate subducting beneath another, and extremely large magnitude events are possible in subduction regions.

Shallow crustal earthquakes include the 1999 M_W 7.62 Chi-Chi Taiwan earthquake, the 2010 M_W 7.0 Darfield (New Zealand) earthquake, the 2010 M_W 7.20 El Mayor-Cucapah earthquake, the 1978 M_W 7.35 Tabas (Iran) earthquake, and the 1999 M_W 7.51 Kocaeli (Turkey) earthquake.

Subduction zones include those generally located in:

- Alaska
- Cascadia (Northwest US and Southwest Canada)
- Central America
- Mexico
- Japan
- New Zealand
- South America (west coast)
- Taiwan

The 2011 M_W 9.12 Tohoku earthquake was a subduction event, as were the 2010 Maule Chile M_W 8.81 earthquake and the 2001 M_W 8.41 South Peru earthquake.

7.2.3 Source-to-Site Distance

Many distance metrics have been proposed for source-to-site distance.

Hypocentral distance, R_{HYP}, is the distance from a particular site to the earthquake hypocenter, the point beneath the surface of the earth where the fault rupture originated.

Epicentral distance, R_{EPI}, measures the distance form a particular site to the surface projection of the hypocenter.

The Joyner-Boore distance, R_{JB}, is the closest distance from a particular site to the surface projection of the entire fault.

R_{RUP} is the closest distance from the site in question to the rupture plane.

Figure 7.1 illustrates several distance measures used in reporting ground motion data during earthquakes.

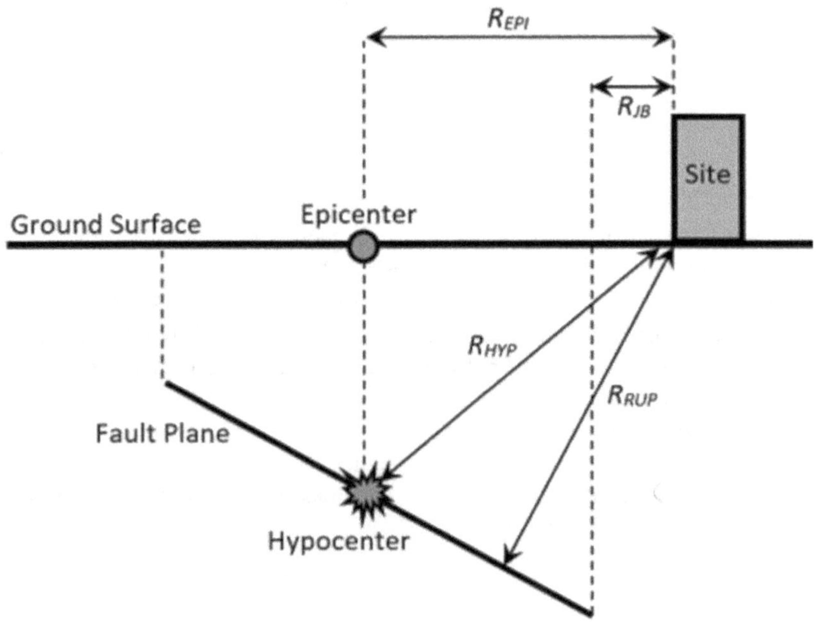

FIGURE 7.1 Source-to-site distance parameters.

7.2.4 Subsurface Properties

Modern codes and specifications define earthquake loading as a design response spectrum. Structures have natural periods of vibration, and the design response spectrum defines the acceleration felt by an elastic structure for any given natural period. So, the *x*-axis of a response spectrum is natural period, and the *y*-axis of a response spectrum is *spectral acceleration* (or, more commonly, *pseudo-spectral-acceleration, PSA*). See Section 7.7.2 for more discussion on PSA. For now, note that mapped, response spectrum accelerations in modern specifications are often for bedrock conditions. A means is necessary, when bedrock is not at the surface, to translate these mapped, bedrock accelerations to the surface, where the structure is located.

To probabilistically estimate effects of ground shaking at a particular site, it is necessary to assess the subsurface conditions at that site. Some of the most frequently employed site characteristics are profile depth to rock, average shear wave velocity (V_{S30}) in the upper 30m (100ft), depth to achieve a shear wave velocity equal to 1 km/s ($Z_{1.0}$), depth to top of rupture (Z_{TOR}), depth to achieve a shear wave velocity equal to 2.5 km/s ($Z_{2.5}$), and Site Class (A, B, C, D, E, F). Profile depth, V_{S30}, and site class are discussed further.

While many design specifications characterize a site based on properties in the upper 30m (presumably, the reason for this is that the Standard Penetration Test provides information in the upper 30m only), the Mississippi Embayment of the New Madrid seismic zone, to give one example, possesses sites with much deeper profiles before rock conditions are encountered, as shown in Figure 7.2 from (Fernandez & Rix, 2006).

This raises the question as to whether code-based site factors, which translate bedrock accelerations to the surface, appropriately characterize the seismic hazard at deep soil sites. Researchers have addressed this issue and two of the most

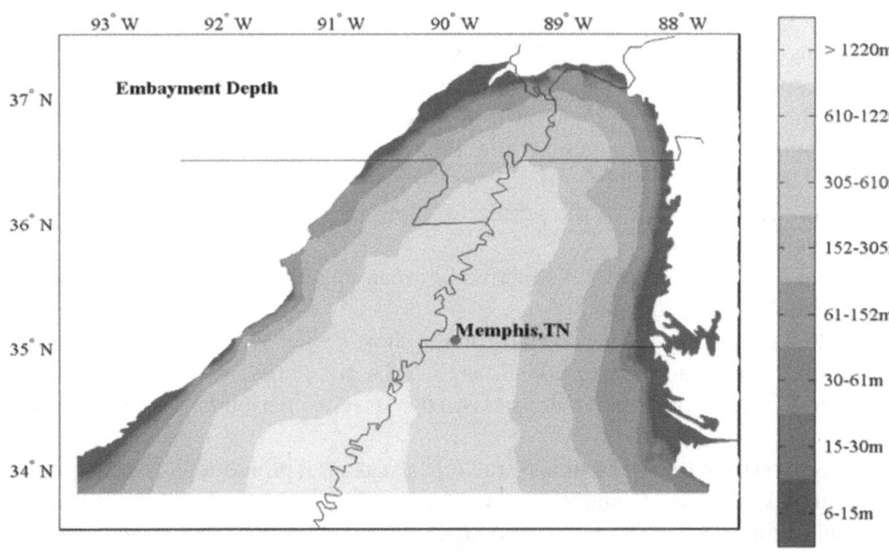

FIGURE 7.2 The Mississippi Embayment (Fernandez, 2007).

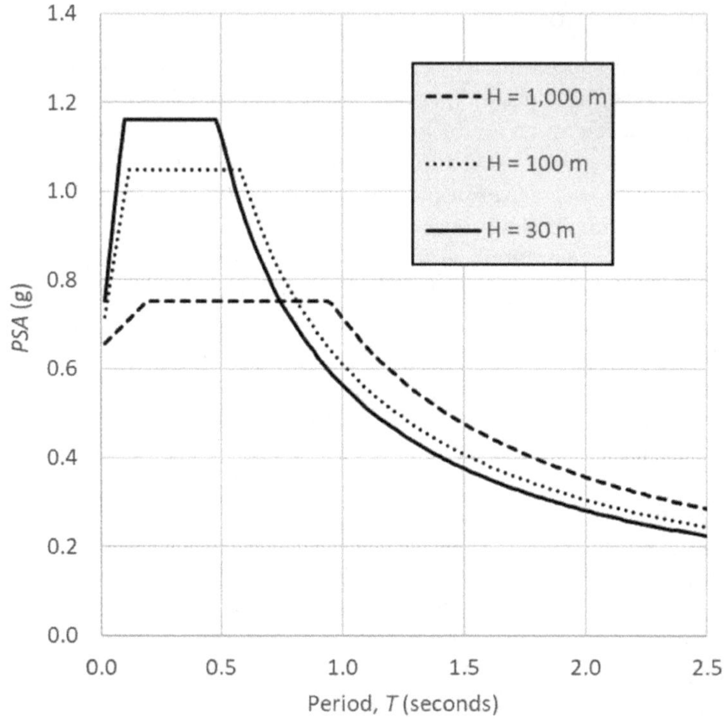

FIGURE 7.3 Effect of profile depth on PSA.

comprehensive studies include (1) that completed at the University of Illinois (Hashash, Tsai, Phillips, & Park, 2008) and (2) that reported in the Korean Society of Civil Engineers (Moon et al., 2016). In general, deep sites tend to exhibit a broader, but lower, spectral acceleration plateau, as depicted in Figure 7.3.

Since the study by Moon et al. (2016) is one of the most recent and detailed, site factors from that study are given here in Tables 7.1 through 7.6. The study included both "lowlands" and "uplands" profiles, with the difference being a lower shear wave velocity in the upper 70 m of the profile for "lowlands" conditions. The subsurface profile depth to rock is *H*.

Refer to Figure 7.45 for the distinction between uplands and lowlands regions of the New Madrid seismic zone.

Note that the site factor, F_a, depends on the short-period PSA for rock conditions, S_S. It has become customary to take S_S as the PSA at a period of 0.20 seconds. The site factor, F_v, on the other hand, depends on the 1-second period *PSA* for rock conditions, S_1.

For deep soil sites, the site factors, F_a and F_v, depend on (1) profile depth, *H*, (2) the level of ground shaking (S_S for F_a and S_1 for F_v), (3) the site class, and (4) shear wave velocities of the subsurface layers. The parameters S_S and S_1 represent mapped, pseudo-spectral accelerations at rock from a hazard analysis at the site. These parameters are defined as

TABLE 7.1
Site Class C Site Coefficient F_a (Moon et al., 2016)

Condition	H (m)	$S_S = 0.25$ g	$S_S = 0.50$ g	$S_S = 0.75$ g	$S_S = 1.00$ g	$S_S \geq 1.25$ g
Uplands	30	1.80	1.75	1.66	1.57	1.54
	100	1.73	1.68	1.59	1.51	1.48
	200	1.67	1.61	1.53	1.46	1.42
	300	1.62	1.56	1.48	1.40	1.37
	500	1.54	1.49	1.40	1.31	1.28
	1,000	1.39	1.32	1.24	1.14	1.11
Lowlands	30	1.72	1.68	1.62	1.49	1.47
	100	1.69	1.65	1.57	1.46	1.43
	200	1.63	1.59	1.51	1.40	1.37
	300	1.58	1.53	1.45	1.34	1.31
	500	1.52	1.46	1.35	1.24	1.21
	1,000	1.38	1.29	1.17	1.07	1.05

TABLE 7.2
Site Class C Site Coefficient F_v (Moon et al., 2016)

Condition	H (m)	$S_1 = 0.10$ g	$S_1 = 0.20$ g	$S_1 = 0.30$ g	$S_1 = 0.40$ g	$S_S \geq 0.50$ g
Uplands	30	1.42	1.40	1.36	1.35	1.34
	100	1.59	1.53	1.48	1.44	1.41
	200	1.82	1.72	1.61	1.55	1.54
	300	1.98	1.83	1.67	1.55	1.54
	500	1.98	1.83	1.67	1.55	1.54
	1,000	1.98	1.83	1.67	1.55	1.54
Lowlands	30	1.40	1.39	1.38	1.29	1.24
	100	1.76	1.73	1.71	1.55	1.51
	200	2.00	1.91	1.80	1.66	1.65
	300	2.22	2.03	1.81	1.66	1.65
	500	2.22	2.03	1.81	1.66	1.65
	1,000	2.22	2.03	1.81	1.66	1.65

TABLE 7.3
Site Class D Site Coefficient F_a (Moon et al., 2016)

Condition	H (m)	$S_S = 0.25$ g	$S_S = 0.50$ g	$S_S = 0.75$ g	$S_S = 1.00$ g	$S_S \geq 1.25$ g
Uplands	30	1.65	1.48	1.21	1.06	0.97
	100	1.54	1.42	1.14	1.00	0.91
	200	1.49	1.39	1.12	0.98	0.88
	300	1.48	1.37	1.09	0.96	0.86
	500	1.46	1.33	1.05	0.92	0.83
	1,000	1.41	1.27	0.97	0.84	0.80

(Continued)

TABLE 7.3 (*Continued*)
Site Class D Site Coefficient F_a (Moon et al., 2016)

Condition	H (m)	$S_S=0.25$ g	$S_S=0.50$ g	$S_S=0.75$ g	$S_S=1.00$ g	$S_S\geq1.25$ g
Lowlands	30	1.45	1.34	1.07	0.94	0.84
	100	1.35	1.24	0.97	0.84	0.75
	200	1.28	1.19	0.92	0.79	0.69
	300	1.25	1.14	0.87	0.75	0.66
	500	1.20	1.09	0.82	0.70	0.64
	1,000	1.15	1.04	0.77	0.66	0.62

TABLE 7.4
Site Class D Site Coefficient F_v (Moon et al., 2016)

Condition	H (m)	$S_1=0.10$ g	$S_1=0.20$ g	$S_1=0.30$ g	$S_1=0.40$ g	$S_S\geq0.50$ g
Uplands	30	2.20	1.87	1.64	1.45	1.34
	100	3.18	2.59	2.17	1.81	1.61
	200	3.60	3.08	2.56	2.19	1.94
	300	3.78	3.29	2.73	2.29	2.02
	500	3.81	3.32	2.76	2.32	2.05
	1,000	3.83	3.34	2.78	2.34	2.07
Lowlands	30	2.30	1.93	1.63	1.38	1.24
	100	3.12	2.58	2.33	1.86	1.60
	200	3.52	3.11	2.66	2.07	1.77
	300	3.76	3.36	2.85	2.15	1.83
	500	3.79	3.39	2.88	2.18	1.86
	1,000	3.81	3.41	2.90	2.20	1.88

TABLE 7.5
Site Class E Site Coefficient F_a (Moon et al., 2016)

Condition	H (m)	$S_S=0.25$ g	$S_S=0.50$ g	$S_S=0.75$ g	$S_S=1.00$ g	$S_S\geq1.25$ g
Uplands	30	1.62	1.10	0.92	0.79	0.69
	100	1.56	1.05	0.87	0.75	0.66
	200	1.52	1.03	0.87	0.75	0.66
	300	1.48	1.03	0.86	0.75	0.66
	500	1.44	1.00	0.83	0.72	0.64
	1,000	1.38	0.94	0.79	0.69	0.62
Lowlands	30	1.46	0.99	0.87	0.74	0.64
	100	1.44	0.95	0.81	0.69	0.58
	200	1.44	0.95	0.81	0.69	0.58
	300	1.42	0.95	0.81	0.69	0.58
	500	1.40	0.95	0.81	0.68	0.58
	1,000	1.34	0.91	0.76	0.63	0.58

TABLE 7.6
Site Class E Site Coefficient F_v (Moon et al., 2016)

Condition	H (m)	$S_1 = 0.10\,g$	$S_1 = 0.20\,g$	$S_1 = 0.30\,g$	$S_1 = 0.40\,g$	$S_S \geq 0.50\,g$
Uplands	30	3.00	2.58	2.15	1.85	1.64
	100	3.90	3.38	2.58	2.27	1.95
	200	4.90	4.06	3.14	2.82	2.32
	300	5.26	4.11	3.14	2.82	2.42
	500	5.26	4.21	3.14	2.84	2.45
	1,000	5.26	4.21	3.14	2.84	2.47
Lowlands	30	3.06	2.36	1.78	1.33	1.08
	100	4.40	3.48	2.77	2.26	1.82
	200	4.80	3.74	3.17	2.63	2.13
	300	5.06	3.93	3.35	2.68	2.33
	500	5.50	4.40	3.48	2.83	2.54
	1,000	5.50	4.52	3.65	2.98	2.65

pseudo-spectral accelerations at periods of 0.20 seconds (S_S) and 1.0 seconds (S_1) and are available from United States Geological Survey data. Once the appropriate site factors have been determined, the pseudo-spectral accelerations at the surface may be obtained from mapped values at rock.

Modern design response spectra are typically either 2-point spectra, with all ordinates controlled by two points, or 22-point spectra. Once the control parameters, S_S and S_1, along with appropriate site factors, F_a and F_v, have been determined, a typical 2-point design response spectrum may be generated. A third mapped parameter, the long-period transition (T_L) is necessary to fully define the spectrum. The entire curve is defined by the design ordinates, S_{DS} and S_{D1}, along with T_L. Figure 7.4 shows the typical 2-point design response spectrum (DRS).

Refer to Sections 7.7.2 and 7.7.3 for additional discussions on response spectra.

Observe from Figure 7.4 that the pseudo-spectral velocity (PSV) is a constant value for all periods between T_S and T_L. This may be deduced from the relationship between PSA and PSV, where PSA is expressed in units of g and PSV is in cm/s. Similarly, for periods greater than T_L, the spectral displacement may be seen to be constant. Equations 7.2 and 7.3 provide the necessary relationships. For these reasons, the corresponding zones of a 2-point DRS are often referred to as the constant velocity and constant displacement regions, respectively:

$$\text{PSV} = (\text{PSA} \times 981)\left(\frac{T}{2\pi}\right) = \left(\frac{S_{D1}}{T} \times 981\right)\left(\frac{T}{2\pi}\right) = 156 S_{D1} \tag{7.2}$$

$$\text{SD} = (\text{PSA} \times 981)\left(\frac{T}{2\pi}\right)^2 = \left(\frac{S_{D1} T_L}{T^2} \times 981\right)\left(\frac{T}{2\pi}\right)^2 = 24.85 T_L S_{D1} \tag{7.3}$$

The average shear wave velocity in the upper 30 m of the soil profile is termed V_{S30}. It is important to recognize that a velocity-based average is not computed the same

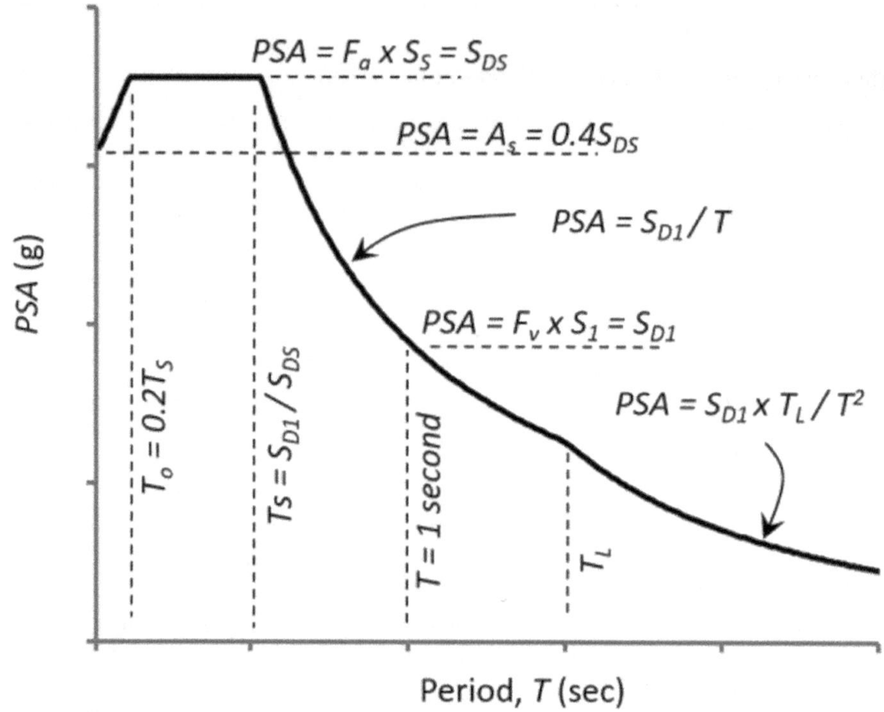

FIGURE 7.4 Typical 2-point design response spectrum (DRS).

way a distance-based average is calculated. The correct calculation of V_{S30} is given in Equations 7.4 and 7.5. The thickness of each individual layer, d_i, and the shear wave velocity for each layer, V_{Si}, are both required for the calculation. Only the top 30 m of the profile is used to calculate V_{S30}. Two calculations are necessary: (1) consider all layers in the top 30 m and (2) consider only the cohesionless layers in the top 30 m:

$$(V_{S30})_1 = \frac{\sum d_i}{\sum \frac{d_i}{V_{Si}}}, \quad \text{all layers} \tag{7.4}$$

$$(V_{S30})_2 = \frac{\sum d_i}{\sum \frac{d_i}{V_{Si}}}, \quad \text{include cohesionless layers ONLY} \tag{7.5}$$

When shear wave velocity measurements are not explicitly made, correlations between shear wave velocity and blow count may be relied upon somewhat. In such cases, it is typically necessary to consider a range of correlated shear wave velocities (0.77 times the correlated velocity to 1.30 times the correlated velocity, for example) to account for uncertainties in the correlations. Inferred shear wave velocities may also give some approximation to shear wave velocity at a site. Table 7.7 is a sampling of correlations available from the literature.

TABLE 7.7
Shear Wave Velocity (m/s) vs. SPT Blow Count

Equation	Region	Soil Type	References
$V_S = 58N^{0.39}$	Turkey	All soils	Dikmen (2009)
$V_S = 73N^{0.33}$	Turkey	Sand	Dikmen (2009)
$V_S = 60N^{0.36}$	Turkey	Silt	Dikmen (2009)
$V_S = 44N^{0.48}$	Turkey	Clay	Dikmen (2009)
$V_S = 142N_{60}^{0.212}$	Peru	Sand	Alhuay-Leon and Trejo-Norena (2021)
$V_S = 39(N_1)_{60}^{0.539}$	Iran	All	Shahgholipour et al. (2024)
$V_S = 98.7N^{0.321}$	Madrid	Sands	Perez-Santisteban et al. (2016)
$V_S = 75.5N^{0.3799}$	Any	All soils	Sil and Haloi (2017)
$V_S = 79.2N^{0.3699}$	Any	Sand	Sil and Haloi (2017)
$V_S = 99.7N^{0.3358}$	Any	Clay	Sil and Haloi (2017)

TABLE 7.8
Hypothetical Profile for V_{S30} Calculation

Layer	d_i, ft	V_{Si}, ft/s	Type
1	12	810	Cohesive
2	12	890	Cohesive
3	48	750	Cohesionless
4	52	985	Cohesionless
5	78	2,750	Cohesionless

N is the uncorrected blow count from the standard penetration test. N_{60} is the blow count corrected for hammer efficiency, borehole diameter, and sample barrel. $(N_1)_{60}$ is further corrected for overburden pressure. The uncorrected blow count, N, has typically been the basis for site classification.

7.2.4.1 Example Problem: Site Classification for Seismic Loading

Suppose a soil profile has been defined, as shown in Table 7.8. Determine V_{S30} for the site.

First, consider all layers in the top 30 m (100 ft):

$$(V_{S30})_1 = \frac{100}{\frac{12}{810} + \frac{12}{890} + \frac{48}{750} + \frac{28}{985}} = 828 \text{ ft/s}$$

Next, consider only the cohesionless layers:

$$(V_{S30})_2 = \frac{76}{\frac{48}{750} + \frac{28}{985}} = 822 \text{ ft/s}$$

So, V_{S30} for this site is 822 ft/s (251 m/s).

While measured shear wave velocity testing is clearly preferred, and at times essential, it is possible to get an idea of the V_{S30} value appropriate for a given site. Correlations between topographic slope and V_{S30} have been a subject of research and the OpenSHA (Field, Jordan, & Cornell, 2003) Java application permits the determination of inferred shear wave velocities, V_{S30}.

Consider, for example a site in Dyersburg, Tennessee – Interstate 155 over the Mississippi River (Caruthersville Bridge). The coordinates of the site are 36.119, −89.615. This is not a plate boundary region but a stable continental region (SCR). The region type in OpenSHA should be selected as *Stable Continent,* and the methodology should be set to *Global* V_{S30} *from Topographic Slope.* With the input shown in Figure 7.5, the inferred V_{S30} is 299 m/s, as shown in Figure 7.6.

ASCE 7-16 (American Society of Civil Engineers, 2017) defines seismic loading for buildings, while AASHTO (AASHTO, 2017) defines seismic loads for bridges. Site classification is similar but not identical for these two specifications. In both cases, a particular project is assigned a Site Class. Determination of the Site Class in ASCE 7-16 is determined in accordance with Table 7.9. The preferred method for Site Class assignment is V_{S30}-based. However, V_{S30} may not be available for many

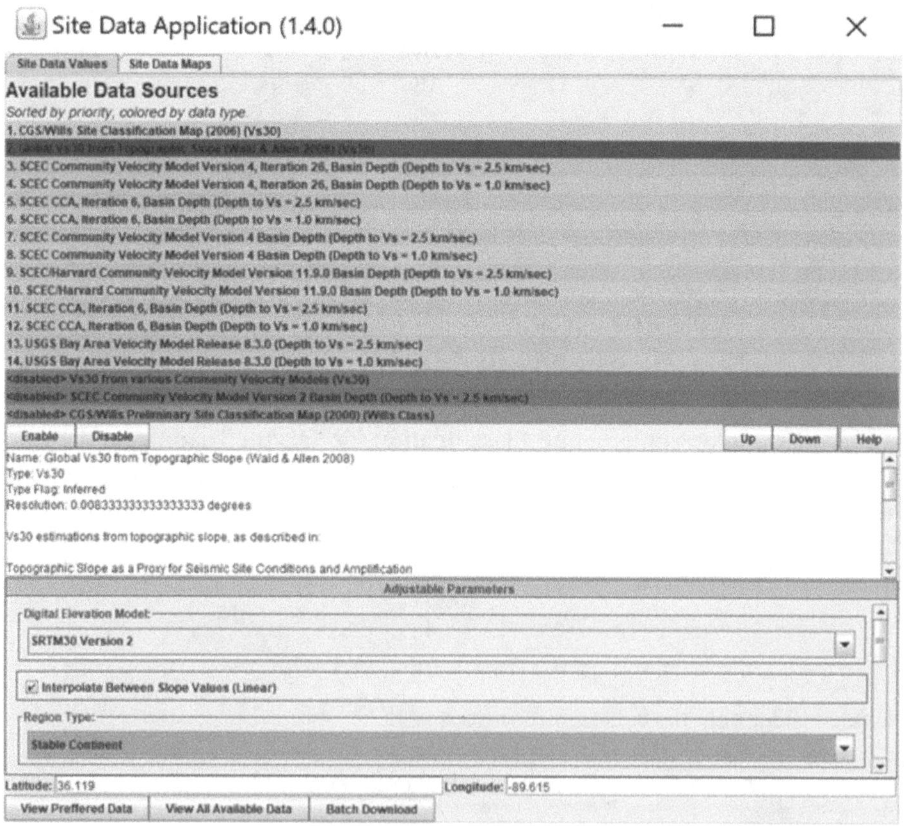

FIGURE 7.5 OpenSHA (Field et al., 2003) inferred shear wave velocity.

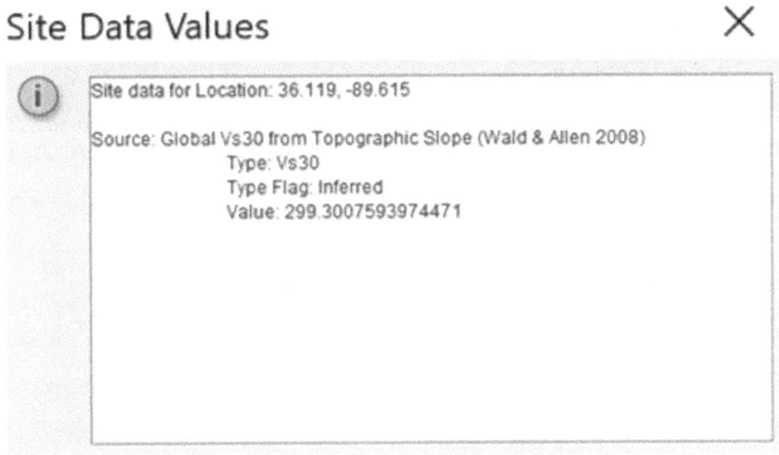

FIGURE 7.6 Inferred shear wave velocity for I-155 over the Mississippi River.

TABLE 7.9
ASCE 7-16 Site Classes

Site Class	V_{S30}, ft/s	N, Blows/ft	S_u, psf
A. Hard rock	>5,000	NA	NA
B. Rock	2,500–5,000	NA	NA
C. Very dense soil and soft rock	1,200–2,500	>50	>2,000
D. Stiff soil	600–1,200	15–50	1,000–2,000
E. Soft clay soil	<600	<15	<1,000
F. Soils requiring site response analysis	Soils vulnerable to liquefaction, highly sensitive clays, peats, highly inorganic clays, high plasticity clays		

projects. So, Site Class determination based on either the standard penetration test or the undrained shear strength is also provided in some specifications. This practice would seem to be somewhat discouraged given the variability in correlations between these properties and V_{S30}. Recall that the SPT gives the blow count (N) required to advance a rod into the ground. The calculation of the average N in the upper 30 m follows the same averaging technique as is used for V_{S30}. ASCE 7-16 Section 20.4.2 requires that N be taken no larger than 100 blows/ft for any layer, including rock layers if encountered in the top 30 m.

ASCE 7-22 uses a refined site classification system as shown in Table 7.10.

ASCE 7-22 further requires that when shear wave velocity in the upper 100 ft is not measured but is estimated using correlations to other measured properties (SPT blow count, shear strength, overburden pressure, CPT resistance, etc.), variability in correlations is to be accounted for. The method requires that site classes based on estimated shear wave velocity is to be derived using V_{S30}, $V_{S30}/1.3$, and $1.3 \times V_{S30}$. If the three estimated velocities result in different site classes, then the most critical site class is to be used at each period.

TABLE 7.10
ASCE 7-22 Site Classes

Site Class	V_{S30} Calculated Using Measured or Estimated V_S Values, ft/s
A. Hard rock	>5,000
B. Medium hard rock	3,000–5,000
BC. Soft rock	2,100–3,000
C. Very dense sand or hard clay	1,450–2,100
CD. Dense sand or very stiff clay	1,000–1,450
D. Medium dense sand or stiff slay	700–1,000
DE. Loose sand or medium stiff clay	500–700
E. Very loose sand or soft clay	<500
F. Soils requiring site response analysis	See ASCE 7-22 Section 20.2.1

7.2.4.2 Example Problem: Design Response Spectrum (2-point)

Suppose that a project is to be constructed on the hypothetical subsurface profile given previously, with $V_{S30}=822$ ft/s and $H=500$ m. The site is an uplands profile in the Mississippi Embayment and the control parameters have been obtained from USGS data at the site as follows:

- $S_S=3.033$ g
- $S_1=0.932$ g
- $T_L=12$ seconds

Using site factors from Tables 7.1 through 7.6, generate a two-point design response spectrum for the project. For this example, the assumption is made that the site factors from (Moon et al., 2016) are applicable.

With $V_{S30}=822$ ft/s, the appropriate site class is D from Table 7.9. With $H=500$ m, from Table 7.3, find $F_a=0.83$. From Table 7.4, find $F_v=2.05$. Determine the control ordinates S_{DS} and S_{D1}, and the control periods, T_S and T_o.

$$S_{DS} = F_a \times S_S = 0.83 \times 3.033 = 2.52 \text{ g}$$

$$S_{D1} = F_v \times S_1 = 2.05 \times 0.932 = 1.91 \text{ g}$$

$$T_S = S_{D1}/S_{DS} = 1.91/2.52 = 0.76 \text{ seconds}$$

$$T_o = 0.2T_S = 0.2 \times 0.76 = 0.15 \text{ seconds}$$

Figure 7.7 shows the 2-point DRS in log-log format.

If the shear wave velocity had not been measured but had been established using correlation to standard penetration test data (N-value blow counts), then a range of shear wave velocities would need to be considered. The low end of this range would

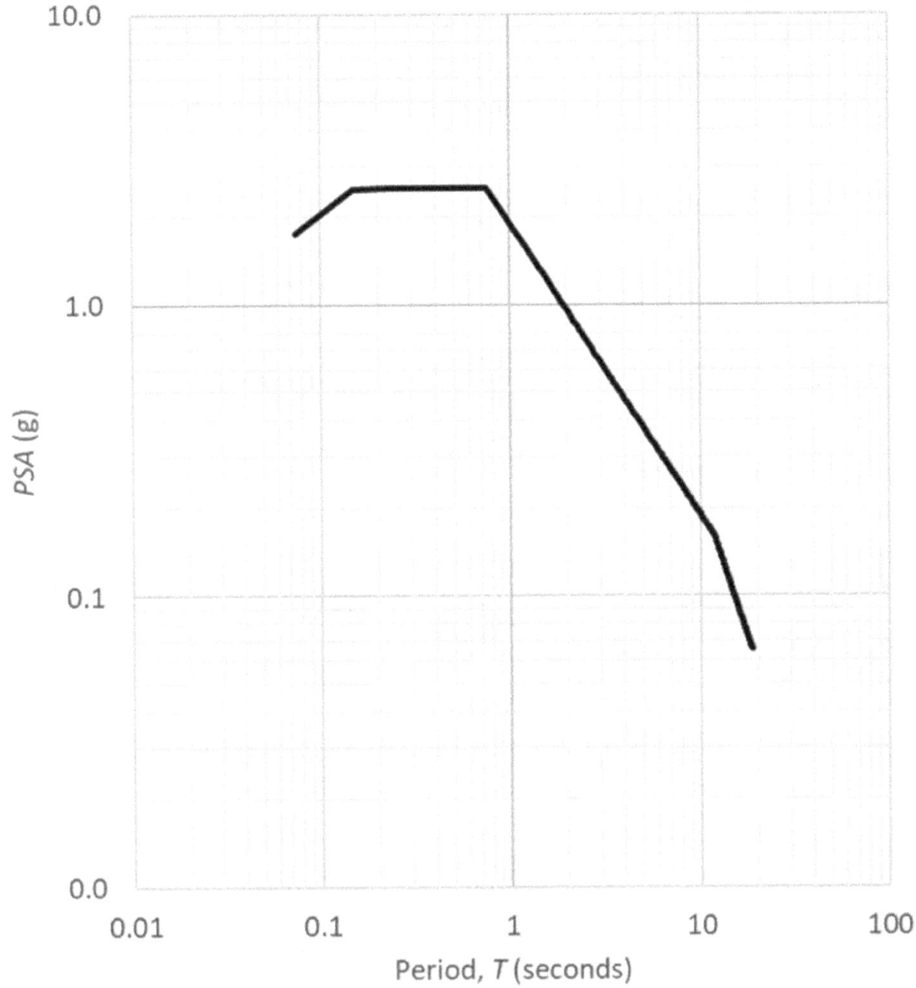

FIGURE 7.7 Example design response spectrum.

be 822/1.3 = 632 ft/s and the upper end of this range would be 822 × 1.3 = 1,069 ft/s. Site classes CD, D, and DE are included in this range of shear wave velocities. At each period of the DRS, the larger site factor and PSA from these three site classes would be taken as the design value at that period.

ASCE 7-22 differs from previous versions of the standard, not only in the definition of refined site classes, but also in the way design response spectra are generated. In ASCE 7-22, design response spectra are defined from 22 points, rather than two points, with interpolation between those points. Site Class is already built into the USGS data for ASCE 7-22 22-point design spectra. No site factors are required unless site-specific values are warranted.

ASCE 7-22 design response spectra are risk-targeted, maximum-direction spectra. AASHTO 2023 design response spectra are risk-targeted, geometric mean spectra.

TABLE 7.11
ASCE 7-16 Site Factor F_a

Site Class	$S_S \leq 0.25$	$S_S = 0.50$	$S_S = 0.75$	$S_S = 1.00$	$S_S = 1.25$	$S_S = 1.50$
A	0.80	0.80	0.80	0.80	0.80	0.80
B	0.90	0.90	0.90	0.90	0.90	0.90
C	1.30	1.30	1.20	1.20	1.20	1.20
D	1.60	1.40	1.20	1.10	1.00	1.00
E	2.40	1.70	1.30	See 11.4.8	See 11.4.8	See 11.4.8
F	See 11.4.8	See 11.4.8	See 11.4.8	See 11.4.8	See 11.4.8	See 11.4.8

TABLE 7.12
ASCE 7-16 Site Factor F_v

Site Class	$S_1 \leq 0.10$	$S_1 = 0.20$	$S_1 = 0.30$	$S_1 = 0.40$	$S_1 = 0.50$	$S_1 \geq 0.60$
A	0.80	0.80	0.80	0.80	0.80	0.80
B	0.80	0.80	0.80	0.80	0.80	0.80
C	1.50	1.50	1.50	1.50	1.50	1.40
D	2.40	2.20[a]	2.00[a]	1.90[a]	1.80[a]	1.70[a]
E	4.20[a]	3.30[a]	2.80[a]	2.40[a]	2.20[a]	2.00[a]
F	See 11.4.8	See 11.4.8	See 11.4.8	See 11.4.8	See 11.4.8	See 11.4.8

a: refer to exception description below

There may be occasions in which ASCE 7-16 site factors are needed. Tables 7.11 and 7.12 provide the site coefficients, F_a and F_v, from ASCE 7-16.

ASCE 7-16 requires a ground motion hazard analysis for the following cases:

- seismically isolated structures and structures with damping systems on sites with $S_1 \geq 0.60$
- structures on Site Class E sites with $S_S \geq 1.00$
- structures on Site Class D and E sites with $S_1 \geq 0.20$

An exception to the requirement for hazard analysis is available in ASCE 7-16 for structures other than isolated structures and structures with damping systems. If the criteria below are incorporated, then the hazard analysis may be waived:

- Site Class E sites with $S_S \geq 1.00$, provided F_a is taken as that for Site Class C.
- Site Class D sites with $S_1 \geq 0.20$, provided $T_S = S_{D1}/S_{DS}$, $T_o = (0.20/1.5) \times (S_{D1}/S_{DS})$, and $F_v = 1.5$ times the tabulated value for Site Class D.
- Site Class E with $S_1 \geq 0.20$, provided $T \leq T_S$ and the equivalent lateral force procedure is used.

7.3 GROUND MOTION CORRECTION AND FILTERING

Ground motions are typically recorded as three perpendicular components: two horizontal and one vertical. So, the ground motion recorded is dependent upon the orientation of the instrumentation. Many times, but not always, instrumentation has been oriented to record North-South and East-West components of ground motion. Ground motion record data virtually always includes orientation of the recording instrument.

When ground motion is recorded during an earthquake, the resulting signal possesses noise and needs to be (1) baseline adjusted and (2) filtered appropriately prior to use in parametric analysis or structural analysis.

7.3.1 Baseline Adjustment

Baseline adjustment removes spurious baseline trends, evident through observation of the integrated displacement history, through least-squares fit of a polynomial with subsequent correction in the recorded accelerogram. Software with baseline adjustment capabilities includes SeismoSignal (SeismoSoft, 2012), PRISM, and TSPP (Boore, 2020) among others.

SeismoSignal is available from https://seismosoft.com/ and PRISM is available from http://sem.inha.ac.kr/prism_2021/index.htm.

TSPP from daveboore.com also provides many useful ground motion processing tools.

7.3.2 Filtering

Filtering involves the removal of certain frequencies from the recorded motion. Filter types include Butterworth, Chebyshev, and Bessel and may be "low-pass," "high-pass," "bandpass," or "band-stop." A "low-pass" filter suppresses frequencies higher than a user-specified value. "High-pass" filters allow only those frequencies higher than a user-specified value. "Bandpass" filters allow only those frequencies within a user-specified frequency range, while "band-stop" filters suppress frequencies within a user-specified range.

Furthermore, filters may be "causal" or "acausal." "Acausal" filtering involves applying a causal filter twice, once forward and once backward, through the ground motion record.

An example of an uncorrected, unfiltered ground motion record is shown in Figures 7.8 (acceleration), 7.9 (velocity), and 7.10 (displacement). The filtered record is shown in Figures 7.11 (acceleration), 7.12 (velocity), and 7.13 (displacement). The record is the East-West component from Station MYG004 recorded during the 2011 M9.12 Tohoku, Japan earthquake.

A few observations in examining the raw and adjusted (corrected and filtered) records will prove valuable.

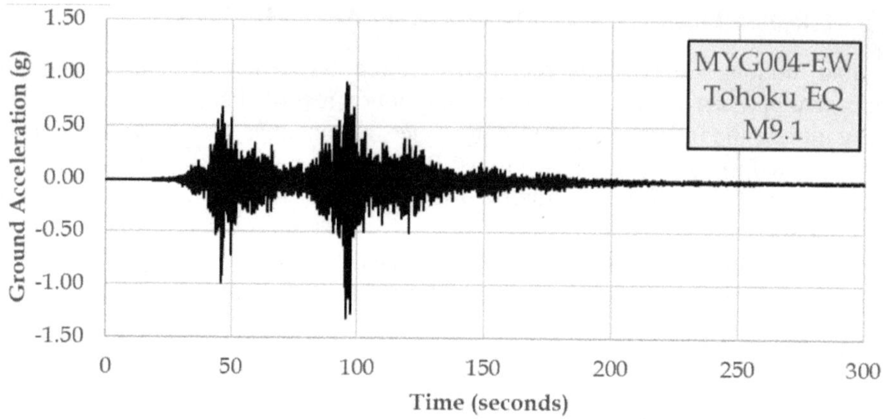

FIGURE 7.8 Uncorrected, unfiltered ground motion record – acceleration.

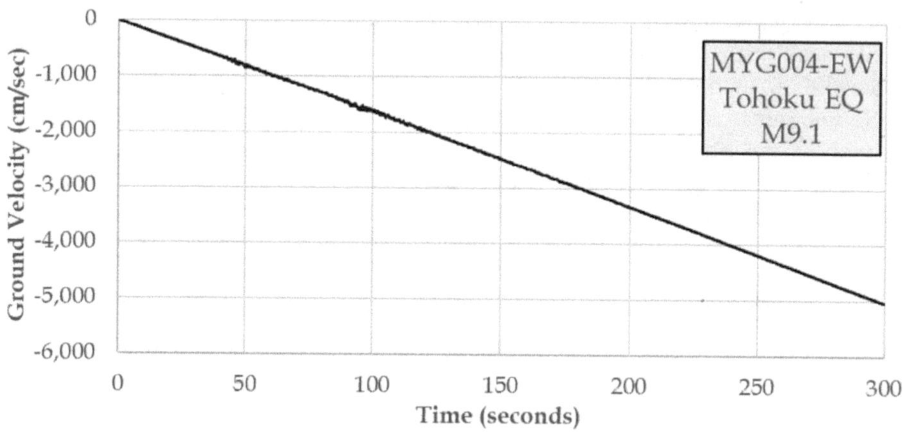

FIGURE 7.9 Uncorrected, unfiltered ground motion record – velocity.

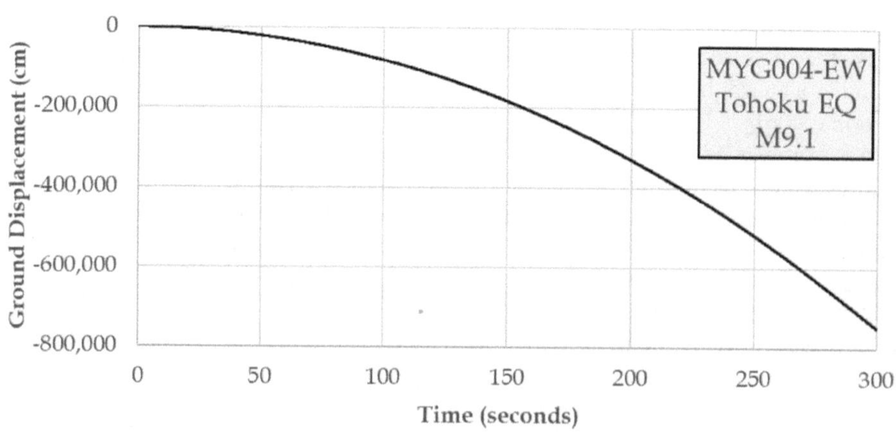

FIGURE 7.10 Uncorrected, unfiltered ground motion record – displacement.

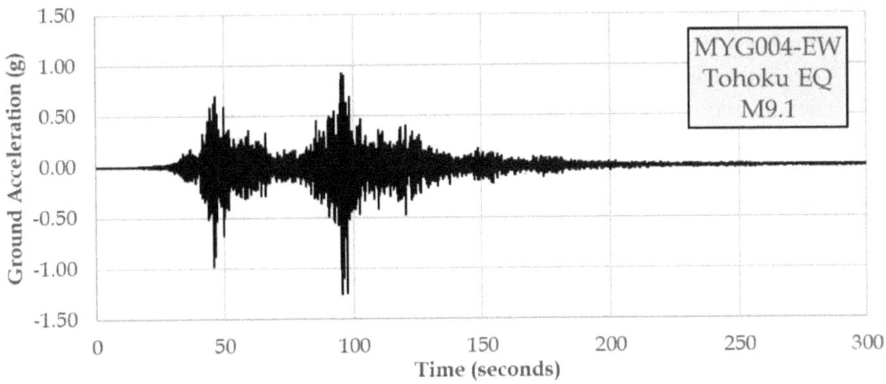

FIGURE 7.11 Corrected and filtered ground motion record – acceleration.

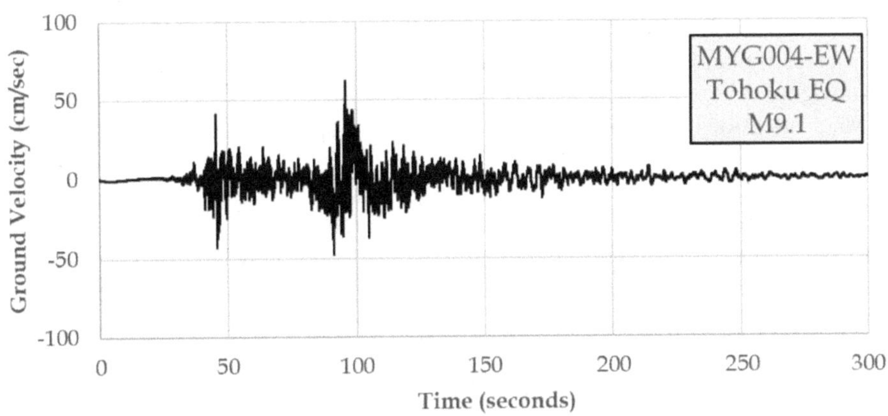

FIGURE 7.12 Corrected and filtered ground motion record – velocity.

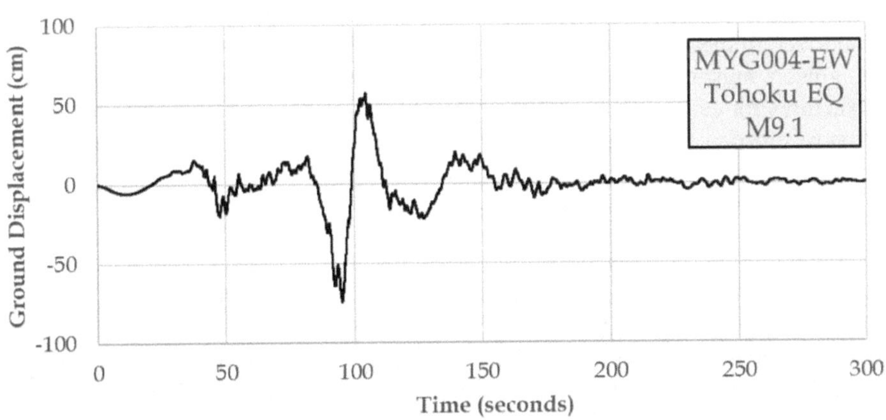

FIGURE 7.13 Corrected and filtered ground motion record – displacement.

- It is difficult to distinguish between the two records, raw and adjusted, by examining the ground acceleration history alone.
- The ground acceleration values for the raw and adjusted records are similar.
- The ground velocity returns to zero for the adjusted record, as would be physically necessary. Unless a permanent ground displacement was observed in field reconnaissance, the same should be true of displacement.
- Integration of the acceleration history to obtain the ground velocity history, and subsequent integration of the ground velocity history to obtain the ground displacement history, are necessary to distinguish the raw and adjusted records.
- Estimating peak ground velocity (PGV) and peak ground displacement (PGD) requires careful attention to baseline adjustment and filtering.

The filtering out of noise and the retention of ground motion is the goal of baseline adjustment and filtering. The range of frequencies specified for filters, the order of the baseline adjustment, and other adjustment/filtering parameters can have a significant impact on the modified ground motion record. Peak ground velocity, PGV, and especially peak ground displacement, PGD, are sensitive to the choice of these parameters.

Much attention has been devoted to the science of appropriately adjusting recorded accelerograms through baseline correction and filtering (Boore & Bommer, 2005; Boore D. M., 2001; Boore & Akkar, 2003).

See Section 7.5.1 for an example which highlights the effect of ground motion processing on the resulting ground motion parameters. It may be difficult to produce accurate values for PGV and PGD as they are highly dependent upon the parameters used for correction and filtering.

7.4 GROUND MOTION DATABASES

The availability of ground motion records for structural analysis and design has grown dramatically over the past two decades. Several databases of record triplets (two horizontal components and one vertical component) and associated metadata from worldwide earthquakes are now available to the engineer. Some of these are summarized in the following sections.

Engineers must make use of properly adjusted and filtered records in the analysis of structures for seismic effects. Otherwise, large displacement drifts in the ground motion record may falsely indicate problems in a structural model, for example.

7.4.1 NGA-West 2 and NGA-East

The original NGA West project was one of the first to provide corrected and filtered ground motion accelerograms for use in structural analysis by response history. Subsequently, expanded versions of the project, NGA-West2 and NGA-East, were completed to enhance the availability of ground motion records from shallow, crustal earthquakes. Both NGA-West2 and NGA-East ground motion databases are located at the Pacific Earthquake Engineering Research Center at the University of California, Berkeley. https://ngawest2.berkeley.edu/

The sparsity of strong motion records from large magnitude events in the Central and Eastern United States (CEUS) renders NGA-East somewhat lacking. When designing for large magnitude strong motion in the CEUS, it may still be necessary to use records from databases other than NGA-East.

NGA-West2 and NGA-East each provide scaling of records to match a target response spectrum and subsequent downloading of metadata (scale factors, event magnitude, subsurface conditions in terms of V_{S30}, etc.) and ground motion records. Options for the target response spectrum basis include geometric mean, *RotD50*, *RotD100* (also known as maximum direction), and SRSS, among many others.

Refer to Section 7.7.1 for further discussion on directionality effects.

7.4.2 NGA Subduction Portal

Over 71,000 record pairs of ground motion data and associated metadata are available from the NGA Subduction Portal. This ground motion database is concerned with subduction events only (Mazzoni, 2022).

https://www.risksciences.ucla.edu/nhr3/nga-subduction

The NGA Subduction Portal facilitates amplitude scaling of subduction ground records to match a target response spectrum, which may be user-defined, generated from ASCE 7 2-point design response spectrum parameters T_L, $S_{DS,}$ and S_{D1}, or generated from a subduction ground motion model. The nature of the target response spectrum must be specified as *RotD50*, *RotD100*, *H*1, *H*2, or vertical. Selection criteria include moment magnitude, average shear wave velocity, rupture distance, significant duration, and peak ground acceleration, among others.

Earthquakes for worldwide subduction zones are included in the NGA Subduction Portal. These subduction zones include:

- Alaska
- Cascadia
- Central America and Mexico
- Japan
- New Zealand
- South America
- Taiwan

Users may also obtain recording station data for subduction events from interactive maps available at the Portal. Figure 7.14 was obtained directly from interactive maps at the NGA Subduction Portal.

Note that the records have been adjusted and filtered, as evidenced by the absence of end-of-record ground velocity and ground displacement drift.

7.4.3 GeoNet

GeoNet is the New Zealand strong motion database. Records from both subduction and shallow crustal earthquakes in the region are available from GeoNet (New Zealand Government, 2017). https://www.geonet.org.nz/data/supplementary/nzsmdb

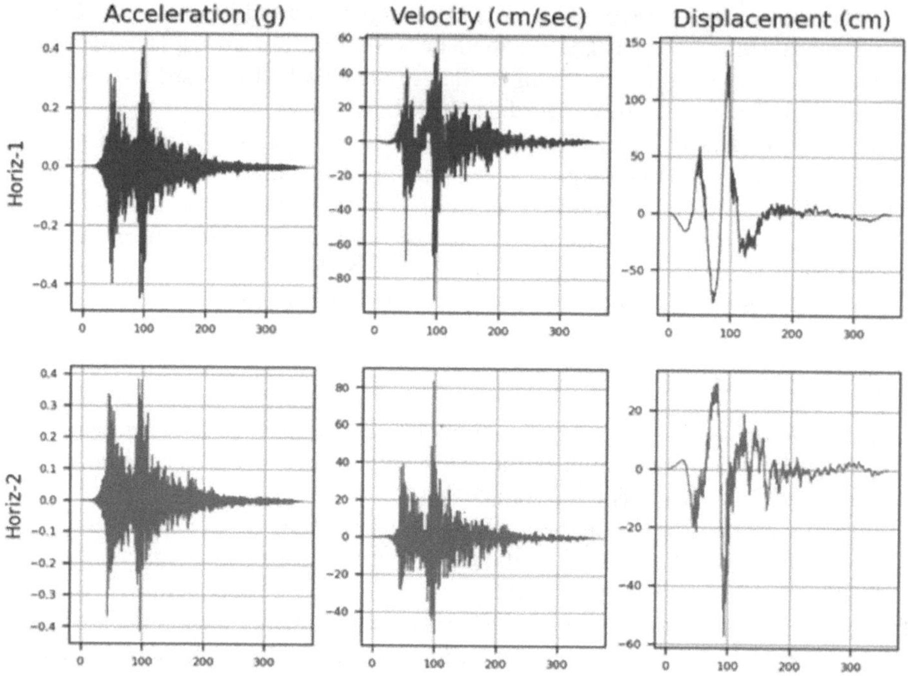

FIGURE 7.14 Station 42212 horizontal ground motion, 2011 Tohoku EQ (M_W 9.12).

Along with flatfiles containing metadata, ground motion records are available for download. While the flatfile includes several thousand records, it is noted on the website that 756 of these three-component accelerograms are appropriate for structural analysis. All records may be downloaded. Or, only the 756 triplets may be downloaded.

7.4.4 Center for Engineering Strong Motion Data

The Center for Engineering Strong Motion Data (US Geological Survey and the California Geological Survey, 2015) houses digital ground motion records dating back to the 1933 Long Beach earthquake. https://www.strongmotioncenter.org/

Both raw and unfiltered records are available for many significant earthquakes, both subduction and shallow crustal. Many convenient features are incorporated into the website. In addition to metadata for the records, Center for Engineering Strong Motion Data (CESMD) provides ground motion plots and sortable tabulated ground motion parameters. Figure 7.15 is a sample from the 2010 Calexico, also known as El Mayor, earthquake.

CESMD includes interactive maps that enable the user to quickly obtain data for available stations. Figure 7.16 is one example, again from the 2010 Calexico earthquake.

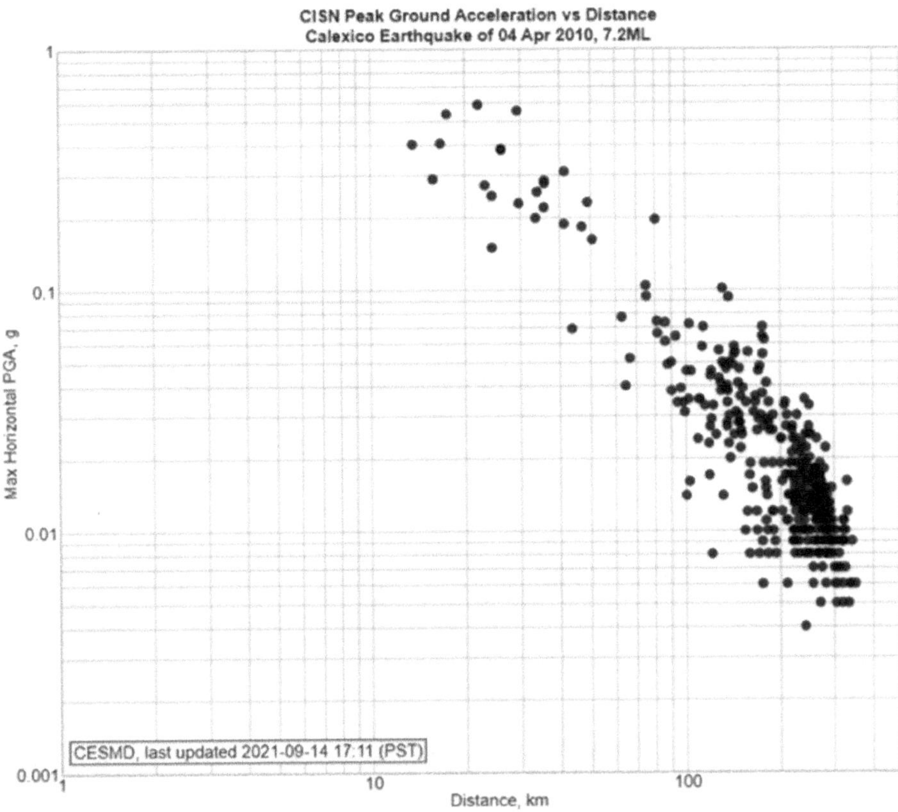

FIGURE 7.15 Sample plot from CESMD for the Calexico Earthquake of 2010.

7.4.5 COSMOS Virtual Data Center

COSMOS is associated with the CESMD. It is described as the global component of the CESMD. Ground motion records from global earthquakes are available; some are corrected and filtered, and some are in raw format. Of particular interest to some users is the link to recent and historical records from the Japanese databases. https://www.strongmotioncenter.org/vdc/scripts/default.plx

7.4.6 TADAS

The Turkish Accelerometric Database and Analysis System (TADAS) provides ground motion data in the form of accelerograms and site characteristic metadata for earthquakes in Turkey and surrounding countries including Iran, Bulgaria, Israel, and Syria (Turkish Ministry of Interior, 2020). https://tadas.afad.gov.tr/map

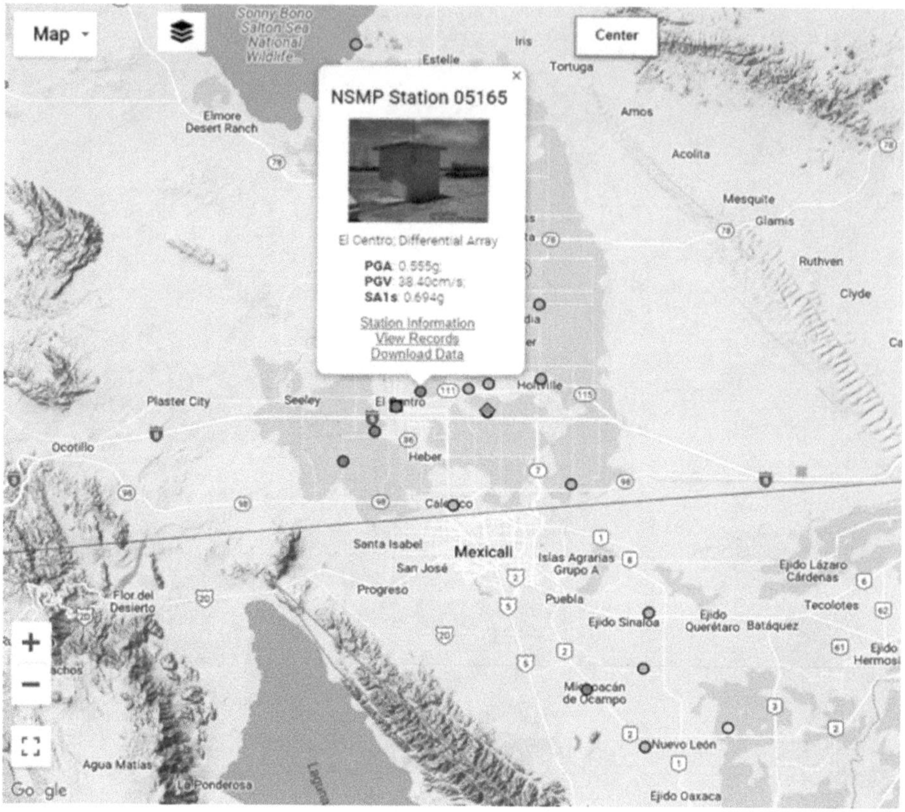

FIGURE 7.16 CESMD interactive map for the 2010 Calexico Earthquake.

Code	Latitude	Longitude	Province	District	PGA(cm/sn2) (N-S) ↓	PGA(cm/sn2) (E-W)
4612	38.02395	36.48187	Kahrama...	Göksun	635.45	523.21
4406	38.34388	37.97378	Malatya	Akçadağ	467.20	409.31
0131	37.856602	36.115347	Adana	Saimbeyli	402.32	331.69
4631	37.966325	37.427653	Kahrama...	Nurhak	337.38	388.61
4409	38.56063	37.49076	Malatya	Darende	287.04	218.04
3802	38.47812	36.50359	Kayseri	Sarız	195.79	220.88
4611	37.7472	37.28426	Kahrama...	Çağlayan...	194.40	139.04

FIGURE 7.17 Search results from the TADAS website.

Figure 7.17 shows a partial tabular result for the February 6, 2023, magnitude 7.6 Elbistan (Kahramanmaras) earthquake in Turkey. Real-time sorting of tabulated results is available. Results in Figure 7.17 have been sorted by descending peak ground acceleration in the north-south direction, for example. Records from the table are downloadable. Both raw data and processed data are available for many records.

7.4.7 ORFEUS

The European Infrastructure for seismic waveform data provides ground motion records and metadata, primarily for events occurring in the European-Mediterranean and Middle-East regions (European Infrastructure for Seismic Waveform Data, 2016). https://esm-db.eu/#/home

Orfeus is a searchable database for ground motion records. Search criteria include magnitude, epicentral distance, peak ground acceleration, peak ground velocity, and date. Also included is a tool for ground motion modification, REXELWEB.

7.4.8 Chile Ground Motion Database

Data and ground motion records for events occurring in Chile are available from the Chile Ground Motion Database. Several events of large magnitude have been included in this database (Civil Engineering Faculty – The University of Chile, 2020). http://terremotos.ing.uchile.cl/

7.4.9 Japan K-NET & Kik-Net

Ground motion records from earthquakes in Japan are available from the website of the National Research Institute for Earth Science and Disaster Resilience: https://www.kyoshin.bosai.go.jp/

User registration is required. Search capabilities enable the user to query only events within a certain magnitude or date range. The records from events beginning in 1996 are maintained up to date.

7.4.10 Engineering Seismology Toolbox

The University of Western Ontario (https://seismotoolbox.ca) provides an online database of (1) stochastically simulated ground motion records for large subduction zone events (magnitude 8.5 and 9.0) characteristic of the Cascadia subduction zone in western Canada, (2) simulated ground motion records compatible with the 2005 National Building Code of Canada for eastern and western Canada, and (3) corrected accelerograms from magnitude 3.5–6.6 earthquakes in southwest Canada occurring between 1994 and 2008.

7.5 GROUND MOTION PARAMETERS

Many intensity measures have been proposed as predictors of earthquake damage potential. Some of those are discussed in detail here.

7.5.1 PGA, PGV, and PGD

Peak ground acceleration (PGA), peak ground velocity (PGV), and peak ground displacement (PGD) are the three most basic ground motion parameters used as indicators of earthquake damage potential.

TABLE 7.13
Effect of Processing on PGA, PGV, and PGD

Parameter	Scheme 1	Scheme 2	Scheme 3
Baseline adjustment	Second order	Second order	Second order
Causal filter type	Butterworth	Butterworth	Butterworth
Filtering	Fourth-order bandpass	Second-order bandpass	Fourth-order bandpass
Frequencies, Hz	0.02–40	0.02–40	0.05–40
PGA, g	1.368	1.383	1.398
PGV, cm/s	95	102	76
PGD, cm	166	214	49

PGA is relatively easily estimable from the recorded ground motion. PGV may be obtained by integrating (numerically) the ground acceleration history, and PGD may be obtained by integrating (numerically) the ground velocity history. Due to noise in ground motion records and the need for baseline adjustment and filtering to remove this noise (see Section 7.3 of this book for additional information), PGV and PGD are relatively difficult to pinpoint. PGV and, particularly, PGD, estimates are highly dependent upon the baseline adjustment and filtering techniques used to remove signal noise.

Take, as an example, the N-S component of the ground motion recorded at the "IBR013" station during the 2011 M_W 9.12 Japan earthquake. Table 7.13 summarizes PGA, PGV, and PGD values obtained using three baseline adjustment and filtering schemes. It is clear from the table that estimated PGA, PGV, and PGD values from ground motion recordings are highly dependent upon the processing scheme applied to the raw record.

7.5.2 Arias Intensity

Arias intensity is a ground motion parameter suggested by Chilean engineer Arturo Arias in 1970 and has been suggested as indicative of landslide potential. Equation 7.6 defines Arias Intensity. Notice that I_A has units of velocity.

$$I_A = \frac{\pi}{2g} \int_0^{t_{max}} [a(t)]^2 \, dt \tag{7.6}$$

7.5.3 Significant Duration

Significant duration is one method used to measure the strong motion duration for a ground motion record. While a typical accelerogram may have a total duration of, say 300 seconds or more, significant duration attempts to define how much of the total accelerogram consists of strong ground shaking.

Significant duration may be D_{5-75}, the time between accumulation of 5% of the total Arias Intensity and accumulation of 75% of the Arias Intensity. Or it may be D_{5-95}, the time between accumulation of 5% of the total Arias Intensity and accumulation of 95% of the Arias Intensity. In fact, significant duration can be computed for any percentage of Arias Intensity accumulation. D_{5-75} and D_{5-95} are the most popular definitions of significant duration.

7.5.4 Bracketed Duration

Bracketed duration is defined as the time between the first and last occurrences of a particular ground acceleration value for a given accelerogram. For example, bracketed duration with a threshold acceleration equal to 0.10 would be the time between the first occurrence of 0.10 g in the accelerograms and the last occurrence of 0.10 g in the accelerograms. As positive and negative accelerations have little physical significance, the bracketed duration is often computed by squaring the ordinates of the accelerograms and searching for the first and last occurrences of the square of the threshold acceleration.

Figure 7.18 shows the Arias Intensity accumulation with time for the North-South component of ground motion recorded at station CHY116 (PEER RSN 1250) during the M_W 7.62 Chi-Chi Taiwan earthquake in 1999. The significant duration, D_{5-95}, is indicated in the figure and is equal to 129 seconds for this record. The bracketed duration is only 0.50 seconds for a threshold acceleration equal to 0.05 g for the same record. So, quite a large difference can exist between significant and bracketed durations. A record possessing high values for both the bracketed duration and the significant duration would seem to be indicative of one having a high damage potential. The bracketed duration depends on the amplitude of ground acceleration while the significant duration depends only on the accumulation of Arias Intensity. Scaling a ground motion record would increase the bracketed duration but leave the significant duration unaltered.

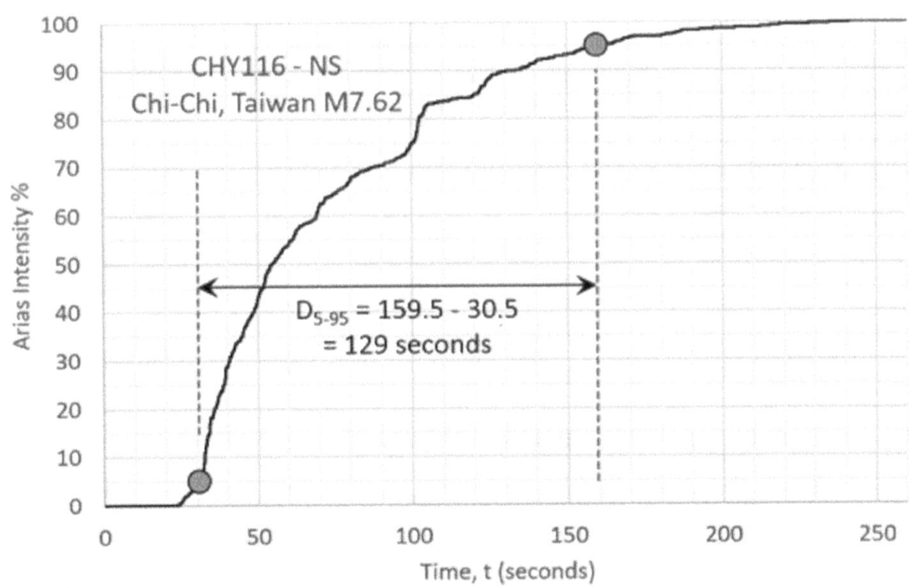

FIGURE 7.18 Arias intensity with D_{5-95} (PEER RSN 1250-NS).

7.5.5 Cumulative Absolute Velocity (CAV)

Cumulative absolute velocity is another ground motion parameter that has been suggested as being indicative of structural damage potential. Notice that CAV has units of velocity and is given by Equation 7.7:

$$\text{CAV} = \int_0^{t_{max}} |a(t)|\, dt \tag{7.7}$$

7.5.6 Specific Energy Density

Specific energy density (SED) is one of the few ground motion parameters computed from the ground velocity history rather than the ground acceleration history. SED has units of length^2 over time. Equation 7.8 may be used to compute SED for a ground motion record:

$$\text{SED} = \int_0^{t_{max}} [v(t)]^2\, dt \tag{7.8}$$

7.5.7 PSV_{MAX}/PSA_{MAX}, PGD/PGV, and SD_{MAX}/PSV_{MAX}

Each of the parameters – PSV_{max}/PSA_{max}, PGD/PGV, and SD_{max}/PSV_{max} – has units of time, usually seconds. These parameters have been proposed as indicative of breakpoints on smoothed design response spectra. While several studies on such estimates may be found in the literature, none seems to be particularly compelling for widespread use in developing design response spectra.

There is evidence of a strong correlation to earthquake magnitude for some of these parameters. Figures 7.19 and 7.20 are based on relationships presented in the literature for subduction earthquakes (Ramon et al., 2024).

Other measures of earthquake ground motion characteristics include the so-called "central period" and "normalized peak ground velocity" (Malhotra, 2006), given by Equations 7.9 and 7.10, respectively. The paper by Malhotra suggests an interesting method to construct smoothed response spectra from ground motion records:

$$T_{cg} = 2\pi\sqrt{\frac{\text{PGD}}{\text{PGA}}} \tag{7.9}$$

$$\overline{\text{PGV}} = \frac{\text{PGV}}{\sqrt{\text{PGD} \times \text{PGA}}} \tag{7.10}$$

7.6 EARTHQUAKE LOADING IN ASCE 7

Earthquake loading in modern design specifications may be in the form of a design response spectrum or a suite of ground motion record pairs appropriately selected and modified for the project site. Both design response spectra and ground motion

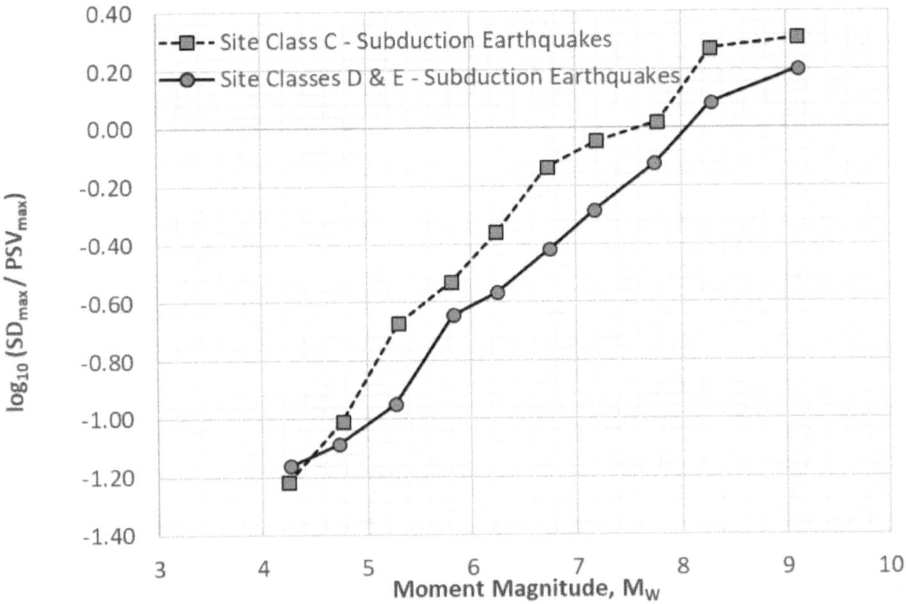

FIGURE 7.19 SD_{max}/PSV_{max} for subduction earthquakes.

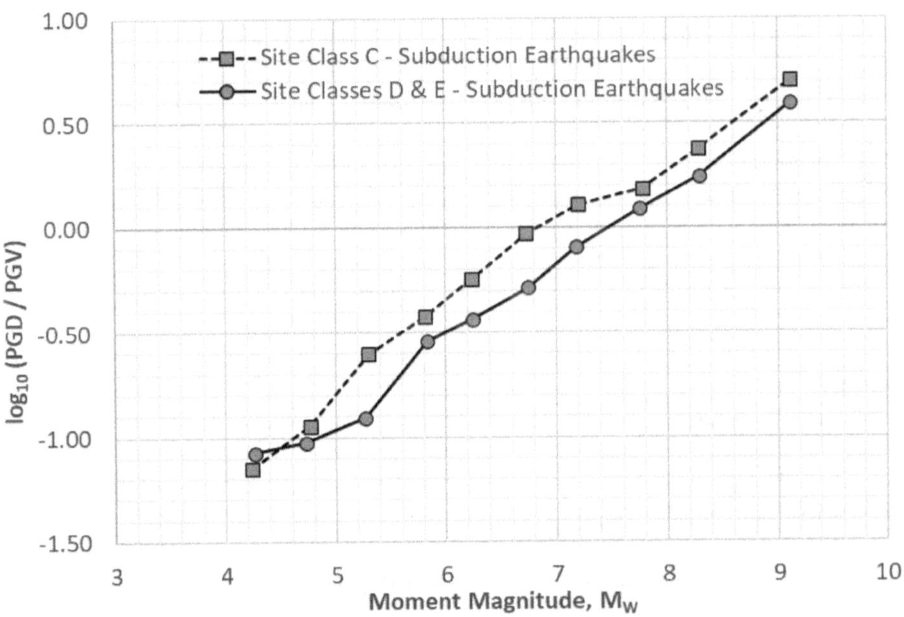

FIGURE 7.20 PGD/PGV for subduction earthquakes.

TABLE 7.14
Site Classification in ASCE 7-16

Site Class	Average Shear Wave Velocity V_{S30}, ft/s
A. Hard rock	>5,000
B. Rock	2,500–5,000
C. Very dense soil and soft rock	1,200–2,500
D. Stiff soil	600–1,200
E. Soft clay soil	<600
F. Soils requiring site response analysis	See ASCE 7-16 Section 20.3.1

record selection/modification criteria are discussed in this section. The concept of design response spectra was introduced in Section 7.2.4.

Typically, design ground motion is defined as having some definite probability of exceedance within a specified period. Ground motion defined thusly across all periods represents a uniform hazard approach to ground motion for design. Both ASCE 7 and AASHTO have moved to risk-targeted ground motion, so that the probability of collapse is more uniform across regions. ASCE 43 used uniform hazard ground motion as a basis but applies period-dependent scale factors to the uniform hazard values.

In ASCE 7-16, design response spectra are based on a 2-point spectrum with site classification as indicated here in Table 7.14.

Design ground motions in ASCE 7-16 (American Society of Civil Engineers, 2017) are risk-targeted, maximum-direction-based (*RotD100*). Distinction is made between MCE_G, the maximum considered earthquake geometric mean acceleration, and MCE_R, the maximum considered earthquake risk-targeted *RotD100* acceleration. MCE_G is not adjusted for targeted risk, it is uniform hazard in nature and represents the geometric mean of two horizontal components. MCE_R is adjusted for targeted risk and is *RotD100* based. MCE_G accelerations are used for liquefaction assessment, lateral spreading, seismic settlement, and other soil-related issues. MCE_R is used in developing design response spectra for structural analysis of buildings and other structures.

Refer to Section 7.7.1 for additional information on directionality of ground motion.

ASCE 7-16 defines near-fault sites as those meeting either of the following:

- The site is within 9.5 miles (15 km) of the surface projection of a known active fault capable of producing $M_W 7$ or larger events.
- The site is within 6.25 miles (10 km) of the surface projection of a known active fault capable of producing $M_W 6$ or larger events.

Near-fault sites are more likely to be subjected to pulse-type ground motions (a noticeable pulse is visible in the velocity history). The PEER Ground Motion database has capabilities for identifying pulse-type ground motion records.

The procedure for generating design response spectra in accordance with ASCE 7-16 is outlined below:

Determine uniform hazard, *RotD100*-based (maximum direction) subsurface, rock spectral accelerations $S_{S\text{-UHS}}$ (spectral acceleration at a period of 0.20 seconds) and $S_{1\text{-UHS}}$ (spectral acceleration at a period of 1.0 seconds). The accelerations should correspond to those having a 2% probability of exceedance in 50 years (about a 2,500-year mean recurrence interval). *GeoMean*-based accelerations may be obtained from the USGS Unified Hazard Tool. *GeoMean*-based accelerations may be converted to *RotD100*-based accelerations by applying factors of (a) 1.1 for periods equal to or less than 0.20 seconds, (b) 1.3 for a period of 1.0 seconds, (c) 1.5 for periods equal to or greater than 5.0 seconds, and (d) linear interpolation for periods between 0.20 and 1.0 seconds and for periods between 1.0 and 5.0 seconds. Refer to ASCE 7-16, Section 21.2 for the above factors.

Apply risk coefficients, C_R, to the accelerations. Risk coefficients are found in ASCE 7-16 Figure 22-18A (C_{RS}) and ASCE 7-16 Figure 22-19A (C_{R1}). This converts uniform hazard accelerations to risk-targeted accelerations. For periods other than 0.20 seconds and 1.0 seconds, risk coefficients are to be determined as follows. At periods equal to or less than 0.20 seconds C_R is equal to C_{RS}. At periods equal to or greater than 1.0 seconds, C_R is equal to C_{R1}. At periods between 0.20 and 1.0 seconds, a linear interpolation is used. The risk-targeted accelerations at 0.20 seconds and 1.0 seconds are S_s and S_1.

Unless site-specific hazard analysis and site response analysis is required, apply site factors from ASCE 7-16, Chapter 11, F_a and F_v. $S_{MS} = F_a \times S_S$ and $S_{M1} = F_v \times S_1$.

Design response spectrum key points are defined by $S_{DS} = 2/3 \times S_{MS}$ and $S_{D1} = 2/3 \times S_{M1}$. Control periods are $T_S = S_{D1}/S_{DS}$, $T_o = 0.20T_S$, and T_L as taken from maps in ASCE 7-16 Chapter 22. Figure 7.21 depicts a generic, 2-point design response spectrum with the control points identified.

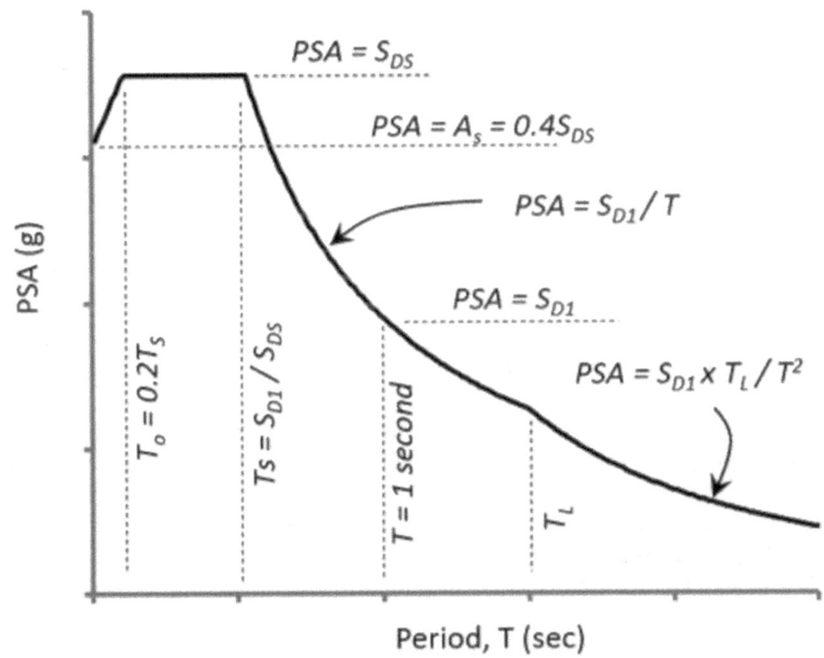

FIGURE 7.21 Generic Design Response Spectrum from ASCE 7-16.

TABLE 7.15
ASCE 7-22 Site Classes

Site Class	V_{S30}, ft/s	V_{S30}, m/s
A. Hard rock	>5,000	>1,500
B. Medium hard rock	3,000–5,000	900–1,500
BC. Soft rock	2,100–3,000	640–900
C. Very dense sand or hard clay	1,450–2,100	440–640
CD. Dense sand or very stiff clay	1,000–1,450	300–440
D. Medium dense sand or stiff clay	700–1,000	210–300
DE. Loose sand or medium stiff clay	500–700	150–210
E. Very loose sand or soft clay	<500	<150
F. Soils requiring site response analysis	See ASCE 7-22 Section 20.2.1	

Ground motion in ASCE 7-22 is based on a 22-point design response spectrum and more site classes. Table 7.15 provides the site classes as assigned in ASCE 7-22. The average shear wave velocity in the upper 100 ft (30 m) of the subsurface profile, V_{S30}, determines the site class for a given location. ASCE 7-22 also specifies a different conversion from *GeoMean* to *RotD100*.

ASCE 7-22 conversion from *GeoMean* to *RotD100* follows the rules listed below. Notice that these more recent conversions are slightly different from those found in ASCE 7-16.

- $\text{PSA}_{\text{RotD100}} = 1.20 \times \text{PSA}_{\text{GeoMean}}$ for period, $T \leq 0.20$ seconds
- $\text{PSA}_{\text{RotD100}} = 1.25 \times \text{PSA}_{\text{GeoMean}}$ for period, $T = 1.00$ seconds
- $\text{PSA}_{\text{RotD100}} = 1.30 \times \text{PSA}_{\text{GeoMean}}$ for period, $T \geq 10.0$ seconds

Both 2-point basis (as adopted in ASCE 7-16 and previous versions of the standard) and 22-point basis (as adopted in ASCE 7-22) design response spectrum data are accessible through USGS web-based tools. https://sites.google.com/site/seismicdesignmaps/

When a 22-point basis DRS is used, there remains the need to define the parameters S_{DS} and S_{D1}. ASCE 7-22 Section 21.4 defines these parameters as follows:

- S_{DS} is equal to 90% of the maximum PSA at any period in the range [0.20–5.00] seconds.
- S_{D1} is equal to 90% of the maximum product, PSA $\times$ T, for periods in the range [1.00–2.00] seconds for sites with average shear wave velocity, $V_{S30} > 1{,}450$ ft/s, and for periods in the range [1.00–5.00] seconds for sites with average shear wave velocity, $V_{S30} \leq 1{,}450$ ft/s.
- S_{D1} is not to be taken less than 100% of the PSA at a period of 1.00 seconds.
- S_{MS} is equal to $1.5 \times S_{\text{DS}}$ and S_{M1} is equal to $1.5 \times S_{D1}$.

The DRS is the loading definition when ASCE 7 permits the use of modal response spectrum analysis in the design of structures. When linear or nonlinear response history analysis (RHA) is employed in design, the DRS is still necessary, but a suite of ground motion records defines the seismic loading for structural design. Ground motion records, when required for response history analysis, must be carefully selected and typically need to be modified. Ground motion modification methods include:

- amplitude scaling
- spectral matching in the time domain by adding wavelets
- spectral matching in the frequency domain by manipulation of Fourier spectra

Refer to Section 7.14.5 for detailed discussions on ground motion selectin and modification.

Linear response history analysis (LRHA) requirements in ASCE 7-22 are found in Section 12.9.2 of the standard. For LRHA, ASCE 7-22 requires no fewer than three pairs of spectrally matched records. The target response spectrum for LRHA is the design response spectrum (DRS), as defined by the 22-point, risk-targeted, *RotD100* values. For those cases in which a 2-point DRS is used in design, the basis for the target is still the risk-targeted, *RotD100* DRS. The DRS is two-thirds of the risk-targeted maximum considered earthquake (MCE_R-based) response spectrum.

- LRHA in ASCE 7-22 requires the definition of a period range of interest for spectral matching of records to the target response spectrum. The period range for LRHA is [0.8 T_{lower}–1.2 T_{upper}].
- T_{lower} is the period of vibration at which no less than 90% mass participation has been achieved in both horizontal directions from a modal analysis of the structure.
- T_{upper} is the larger of the fundamental (first mode) periods in the two horizontal directions from a modal analysis of the structure.
- For LRHA, ASCE 7-22 requires that the average of the 5% damped PSA ordinates of the spectrally matched records be within 10% of the target response spectrum at all periods within the period range of interest.
- Strictly speaking, ASCE 7-22 requires spectrally matched, as opposed to amplitude-scaled, records when LRHA is adopted in the design process.

For nonlinear response history analysis (NLRHA) ASCE 7-22 devotes the entirety of Chapter 16 to requirements for such procedures in the design of structures.

- No fewer than 11 record pairs are required for NLRHA.
- The target response spectrum for NLRHA is the maximum-direction (*RotD100*), risk-targeted, maximum considered earthquake (MCE_R) response spectrum, not the design response spectrum.
- Ground motion candidate records for the suite of records must be selected from events "within the same general tectonic regime and having generally consistent magnitudes and fault distances as those controlling the target spectrum". Additionally, candidate records shall have "a spectral shape similar to that of the target".

- Simulated records are permissible when records from recorded events meeting the requirements are not available.
- Records may be either amplitude-scaled or spectrally matched. Spectral matching is not permissible for near-fault sites unless the pulse character of the ground motion is retained in the matching process.
- For amplitude scaling, both components of each record receive the same scale factor. Scale factors are to be computed so that the average of the *RotD100* spectra form all records in the suite "generally matches or exceeds the target" over the period range of interest. The average for all records shall not fall below 90% of the target at any period within the period range of interest.
- For spectral matching, each record pair is modified such that the average of the *RotD100* spectra for the records equals or exceeds 110% of the target over the period range of interest.
- The upper bound for the period range of interest is equal to twice the largest fundamental period in the two horizontal directions from a modal analysis of the structure.
- The lower bound for the period range of interest is the lesser of (1) the period at which 90% mass participation in both horizontal directions is achieved from a modal analysis of the structure and (2) 20% of the smaller of the fundamental periods in the two horizontal directions.
- ASCE 7-22 Section 21.2.1 specifies a natural logarithm-based standard deviation equal to 0.6 across all periods of interest for site-specific ground motion development. It is not clear whether the natural logarithm-based standard deviation of the record suite is to have a target value of 0.60 across all periods within the period range of interest in ground motion selection and modification.

For an elastic-perfectly-plastic (EPP) system, it could be argued that the upper bound on the period range of interest should be extended beyond twice the largest fundamental period. The effective period for an EPP at a displacement ductility equal to 6 is $T_i \times (6)^{1/2} = 2.45$. An upper bound on the period range of interest for such systems might better be taken as 2.5 times the largest fundamental period based on this argument.

The commentary to ASCE 7-22, Section C16.2.2, notes that when spectral shape is considered in ground motion selection, the magnitude, distance, and site class range requirements may be relaxed. Match to spectral shape has been identified as the single most important factor in the selection of ground motion records for design (NEHRP Consultants Joint Venture, 2011).

Two criteria which measure match to spectral shape are the mean-square-error, MSE, and the D_{RMS} parameters. Equation 7.11 gives an expression for the MSE. Equation 7.12 defines the DRMS parameter. The computation of a scale factor, *f*, which minimizes the MSE between a scaled record and the target response spectrum is given in Equation 7.13. The pre-scaled MSE may be computed using a scale factor of 1.0 in Equation 7.11. However, the appropriate measure of match to spectral shape is the post-scaled MSE:

$$\text{MSE} = \frac{\sum w(T_i) \cdot \left\{ \ln\left[\text{PSA}_{\text{TAR}}(T_i)\right] - \ln\left[f \cdot \text{PSA}_{\text{GM}}(T_i)\right] \right\}^2}{\sum w(T_i)} \quad (7.11)$$

$$D_{\text{RMS}} = \frac{1}{N}\sqrt{\sum_{i=1}^{N}\left(\frac{(\text{PSA}_{\text{GM}})_{\text{Ti}}}{\text{PGA}_{\text{GM}}} - \frac{(\text{PSA}_{\text{TAR}})_{\text{Ti}}}{\text{PGA}_{\text{TAR}}}\right)^2} \quad (7.12)$$

$$\ln f = \frac{\sum \left[w(T_i) \cdot \ln \frac{\text{PSA}_{\text{TAR}}(T_i)}{\text{PSA}_{\text{GM}}(T_i)} \right]}{\sum w(T_i)} \quad (7.13)$$

- PSA_{GM} is the pseudo-spectral acceleration for the ground motion record.
- SA_{TAR} is the pseudo-spectral acceleration for the target response spectrum.
- PGA_{GM} is the peak ground acceleration for the ground motion.
- PGA_{TAR} is the peak ground acceleration for the target response spectrum.
- w is a weighting factor which may be used to assign greater importance to certain periods in the period range of interest, if desired. Typically, the weights for each period within the range of interest are assigned a value of 1.00.

Depiction of ground motion suite results like that shown in Figure 7.22 is valuable in demonstrating the adequacy of a ground motion suite to satisfy various criteria. In a single plot, Figure 7.22 portrays the minimum (0.935), average (1.000), and maximum (1.028) suite-mean-to-target PSA ratios over a period range of interest equal to [0.2–3.0] seconds.

7.6.1 Example Problem: Spectral Matching for LRHA in ASCE 7-22

ASCE 7-22 requires that ground motions for linear response history analysis be spectrally matched to the target response spectrum. The target spectrum is the design response spectrum (DRS), either a 22-point basis DRS or a 2-point DRS defined by S_{DS} and S_{D1}. If site-specific analysis is conducted for a site, then the target response spectrum for the site-specific results is used.

The period range of interest is defined as 0.8 T_{lower} to 1.2 T_{upper}. T_{lower} is the period at which at least 90% of the mass has been accounted for in both horizontal directions. T_{upper} is the larger of the two fundamental periods in the two horizontal, orthogonal directions.

According to ASCE 7-22, each component is to be matched to the target response spectrum such that the average of the suite response spectra does not deviate from the target by more than 10% at any period within the range of interest in either direction.

For this example, a hypothetical hospital project near downtown Martin, Tennessee, is to be designed using linear response history analysis. The project coordinates are 36.343 N, 88.836 W (36.343, −88.836). For this hypothetical

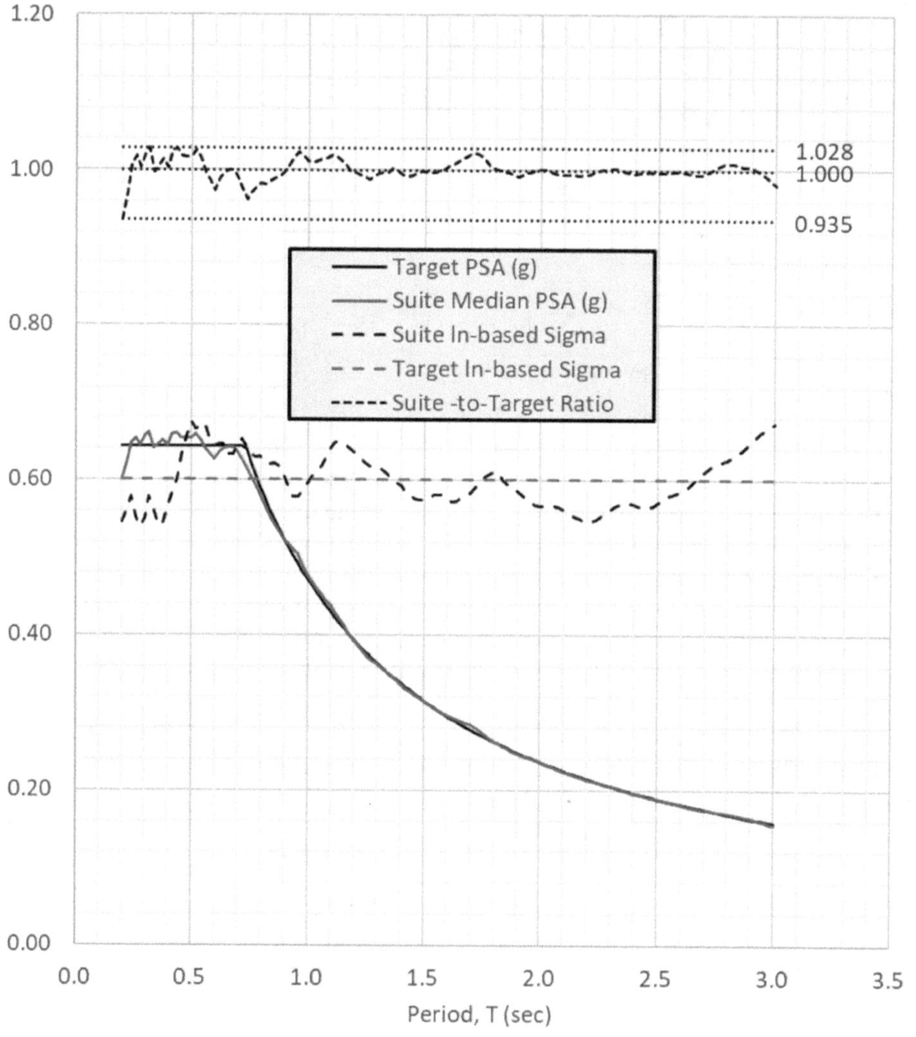

FIGURE 7.22 Ground motion suite mean vs. target PSA.

project, assume that shear wave velocities have been measured and the conditions warrant a Site Class D assignment.

Table 7.16 provides the Risk Category IV (appropriate for a hospital) design response spectrum obtained from the online ASCE 7 Hazard Tool (American Society of Civil Engineers, 2024). Both 22-point and 2-point DRS are shown in Table 7.16.

The target response spectrum is a maximum-direction (*RotD100*) spectrum. Normally, in ground motion modification, the *RotD100* of a ground motion record pair would be matched to the *RotD100* target. However, for linear response history analysis, ASCE 7-22 specifies that each component is to be matched to the target. So, any software capable of matching a single component to a target response spectrum would be a viable option.

TABLE 7.16
Target Response Spectrum for Spectral Matching Example

Period, *T* (seconds)	22-Point PSA (g)	Period, *T* (seconds)	2-Point PSA (g)
0.000	0.460	0.000	0.330
0.010	0.470	0.010	0.368
0.020	0.700	0.020	0.407
0.030	0.790	0.030	0.445
0.050	0.890	0.050	0.522
0.075	0.890	0.075	0.618
0.100	0.900	0.100	0.715
0.150	0.850	0.130	0.830
0.200	0.930	0.200	0.830
0.250	0.860	0.250	0.830
0.300	0.780	0.300	0.830
0.400	0.720	0.400	0.830
0.500	0.680	0.500	0.830
0.750	0.620	0.650	0.830
1.000	0.540	1.000	0.540
1.500	0.380	1.500	0.360
2.000	0.270	2.000	0.270
3.000	0.160	3.000	0.180
4.000	0.110	4.000	0.135
5.000	0.081	5.000	0.108
7.500	0.049	7.500	0.072
10.00	0.031	10.00	0.054

In cases where the *RotD100* of ground motion record pairs is to be matched to a *RotD100* target, the Python-based software REQPY (Montejo, 2021) is freely available. In fact, REQPY has the capability of matching the *RotDnn* spectrum of ground motion pairs to a target *RotDnn* spectrum, where *nn* is any desired value between 00 and 100.

In cases where individual components of a ground motion record pair are to be spectrally matched to a target response spectrum, SeismoMatch (SeismoSoft, 2010) is one available tool. SeismoMatch provides spectral matching capabilities in the time domain by the addition of wavelets to a record. Spectral matching in the frequency domain, wherein the Fourier spectrum of a ground motion record is manipulated and converted to a new record, is available in SeismoArtif (SeismoSoft, 2012).

In cases where amplitude scaling of ground motion record pairs such that the geometric mean response spectrum of the two components in the pair is scaled to a geometric mean-based target, SigmaSpectra (Kottke & Rathje, 2012) is a valuable tool. SigmaSpectra permits the definition of a target response spectrum and a target natural-logarithm-based variability.

For spectral matching of ground motion to be used for linear response history analysis, SeismoMatch is one excellent option and will be used here for time-domain-based spectral matching. For amplitude scaling, SigmaSpectra is one excellent option and will be used in this text for ground motion modification by amplitude scaling, applicable to nonlinear response history analysis.

While no detailed criteria for candidate record pairs are defined in ASCE 7-22 for linear response history analysis, for this example, only records from magnitude 7–8 events recorded on Site Class D subsurface conditions will be used as seed records to be spectrally matched to the target response spectrum. This issue would seem to be much less important for linear compared to nonlinear analysis.

Spectral matching, whether time-domain-based or frequency-domain-based, will generally be more effective if the ground motion records have a spectral shape that reasonably well matches the shape of the target spectrum.

Suppose that a period range of interest has been determined from modal analysis to be [0.30–2.40] seconds. Based on a post-scaled match to the 2-point DRS spectral shape (Equations 7.11 and 7.13 to minimize the mean-square-error), three seed records from Site Class D subsurface conditions are selected:

1. 1999 Chi-Chi, Taiwan, M_W=7.62, Station TCU038, Scale Factor=2.043
2. 2010 El Mayor-Cucapah, Mexico, M_W=7.20, Station Westside Elementary School, Scale Factor=1.392
3. 1986 Taiwan SMART1(45), M_W=7.30, Station SMART1 O03, Scale Factor=1.927

Scale factors have been determined solely to potentially improve the spectral matching. Spectral matching in the time domain with SeismoMatch was used to perform the matching process.

Figure 7.23 shows the original spectra of the scaled ground motions along with the 3-record mean and the target for the first component and for the second component. Figure 7.24 shows the corresponding plot for the spectrally matched records. For component 1, the minimum and maximum ratios of suite-mean-to-target PSA are 0.984 and 1.041, respectively. Both values are well within the permissible range of 0.90–1.10. For component 2, the minimum and maximum ratios are 0.979 and 1.044. Components 1 and 2 were arbitrarily selected from each record.

Figure 7.25 shows the original and spectrally matched acceleration, velocity, and displacement histories for the N/S component of the SMART1(45) record. It is important to inspect not only the acceleration histories of spectrally matched records, but also the integrated velocity and displacement histories, to ensure realistic records have been produced from the matching process. The Figures represent the results obtained when using the 2-point design response spectrum for the target. Successful matching was performed using the 22-point target as well.

7.7 GROUND MOTION RESPONSE SPECTRA

Prior to a further discussion on ground motion response spectra, it is important to understand the various definitions for design ground motion with respect to directionality.

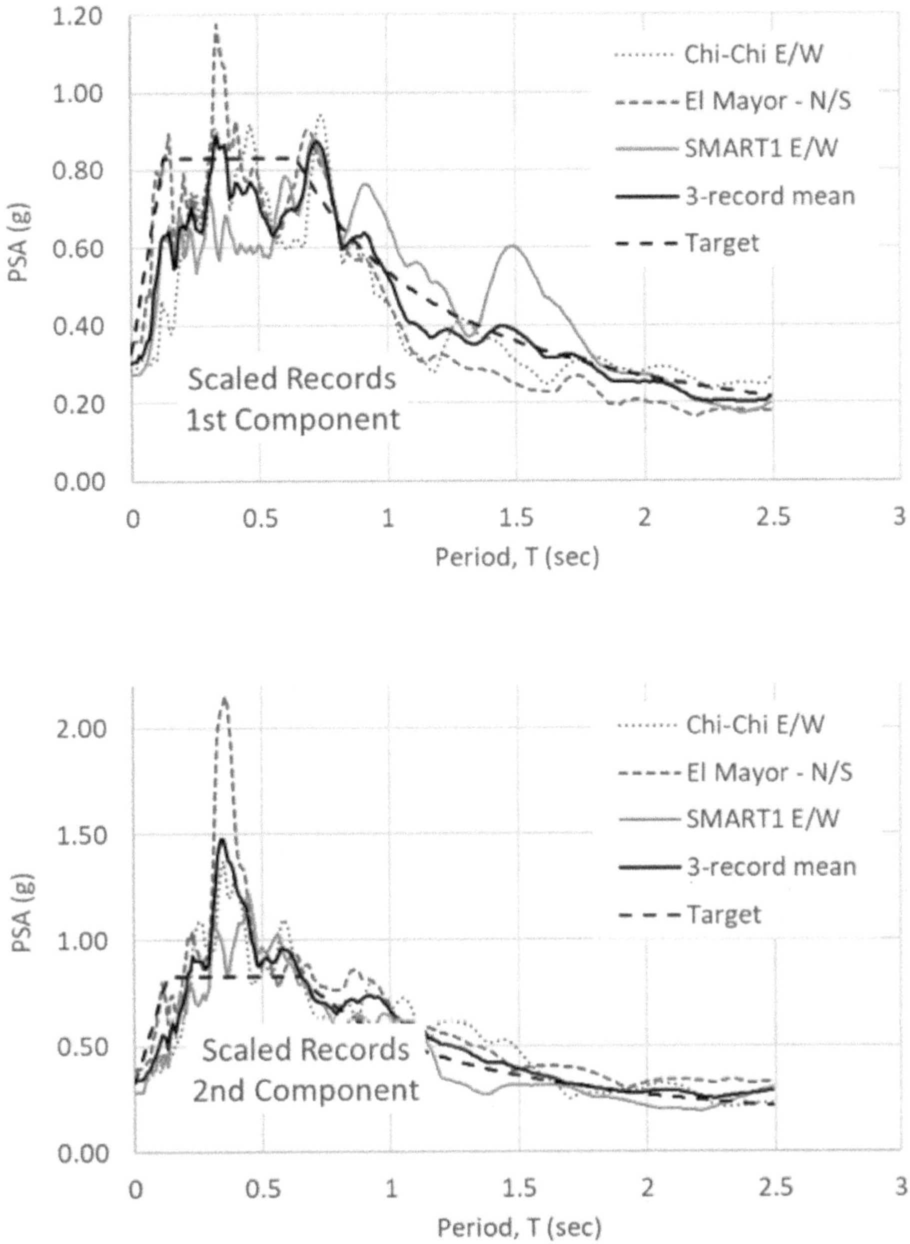

FIGURE 7.23 Scaled and target response spectra.

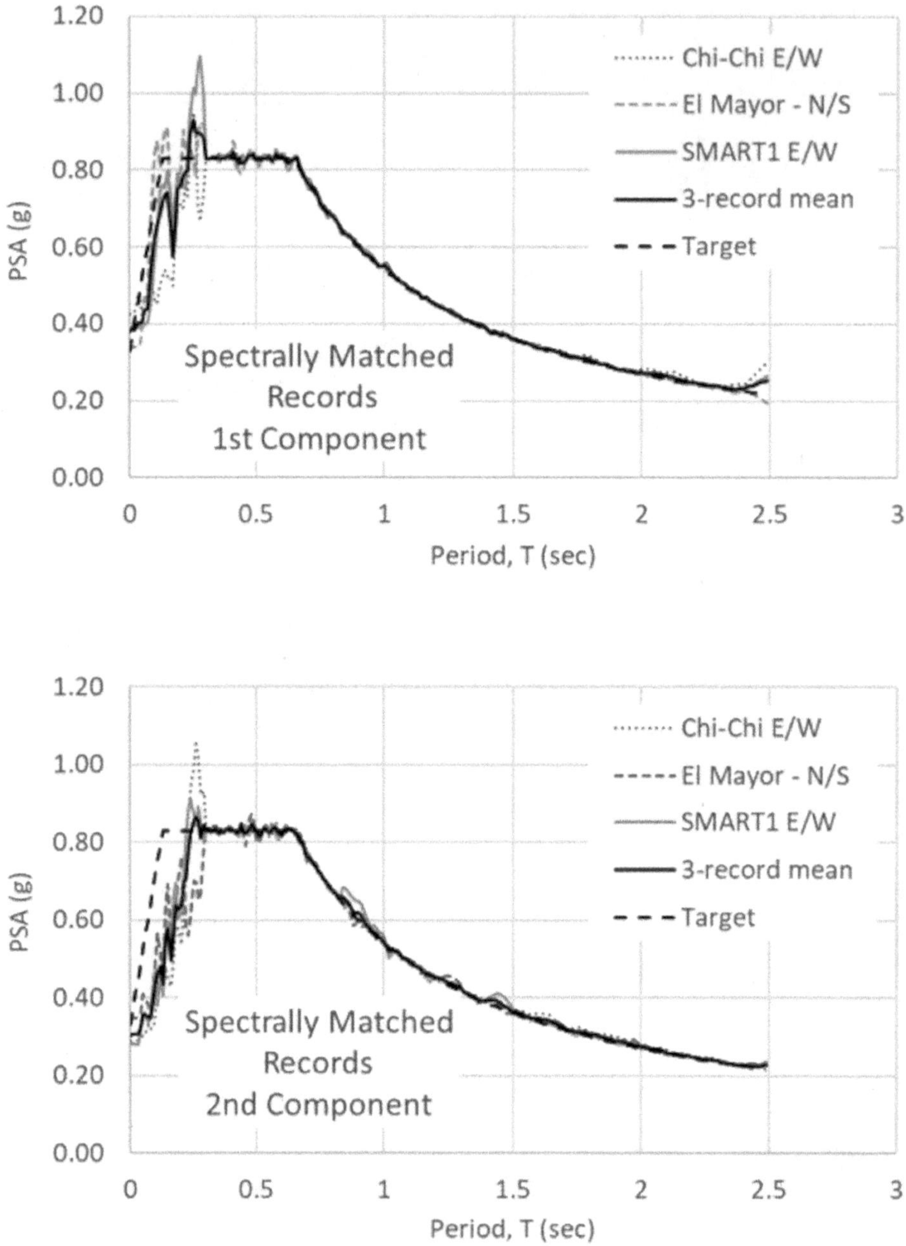

FIGURE 7.24 Spectrally matched response spectra.

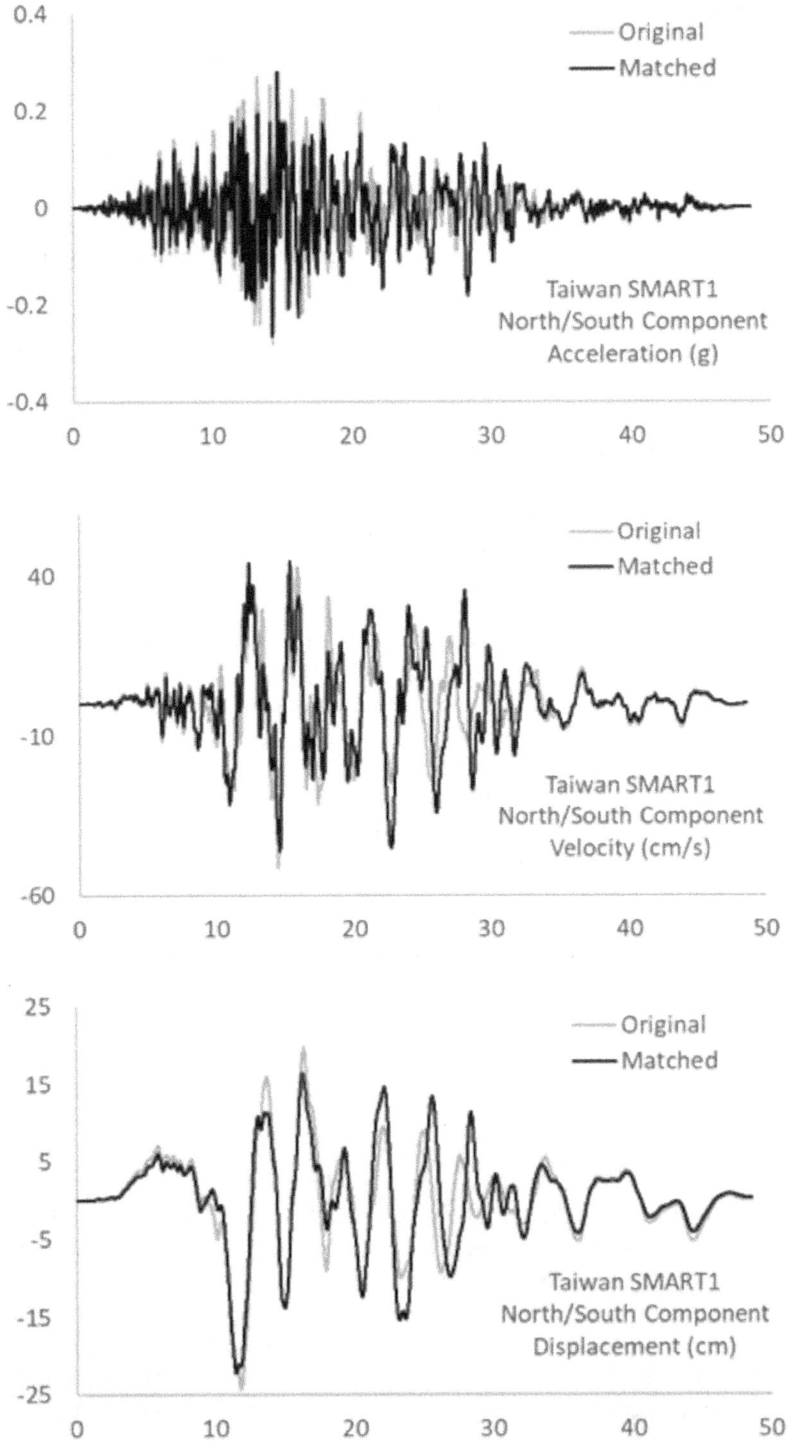

FIGURE 7.25 Taiwan SMART1(45) N/S spectrally matched ground histories

7.7.1 Ground Motion Directionality

The orientation of ground motion recording instrumentation affects the accelerograms obtained by the instrumentation during strong shaking. Three components are typically recorded – two orthogonal horizontal components and one vertical component. Design ground motions have historically been represented by the geometric mean of the two horizontal components. Recent design ground motion in multiple specifications has moved to a maximum-direction (*RotD100*) basis. There are other methods used to measure the horizontal ground shaking intensity in a single parameter.

Most of the original ground motion models were designed to predict spectral acceleration values corresponding to the geometric mean (GM) of two horizontal components. Historical design response spectra, though not explicitly defined as such, were also based on the geometric mean. For each period of vibration, spectra are computed for the two as-recorded horizontal ground motion components, PSA_{H1} and PSA_{H2}. The design spectral ordinate is then given by Equation 7.14 according to the definition of the geometric mean.

$$\mathrm{PSA_{GM}} = \sqrt{\mathrm{PSA}_{H1} \cdot \mathrm{PSA}_{H2}} \tag{7.14}$$

If instrumentation had been oriented differently, then two different ground motion records would have been obtained during strong ground shaking. For horizontal ground motion recorded at the Yermo fire station during the 1992 M_W7.28 Landers earthquake, Figures 7.26, 7.27, and 7.28 show the variation of peak ground acceleration, velocity, and displacement, respectively, with angle of rotation, θ, as an example of the directionality effect. The response spectra of the recorded ground motion vary as well. Plots similar to Figure 7.26 could be developed for any ground motion parameter. For any rotation angle, θ, the ground motions obtained can be expressed as given by Equations 7.15 and 7.16. The rotated components may be computed for each non-redundant rotation angle and the geometric mean spectrum may be computed. The envelope of all thusly obtained spectra is *GMRotD100*. The median of all thusly obtained spectra is *GMRotD50*:

$$a_1(\theta, t) = a_X(t)\cos\theta + a_Y(t)\sin\theta \tag{7.15}$$

$$a_2(\theta, t) = a_X(t)\sin\theta + a_Y(t)\cos\theta \tag{7.16}$$

More recently, design specifications (namely, ASCE 7-16 and ASCE 7-22), have specified design spectra to be based on the maximum horizontal direction (*RotD100*), as previously depicted in Figure 7.26 for PGA of the Landers record pair. This ground motion definition is not a geometric mean, but also relies upon rotation of ground motion components. A single horizontal component is computed according to Equation 7.17 for each rotation angle. Spectra are computed for that component across all periods and for all non-redundant rotation angles (up to 180°).

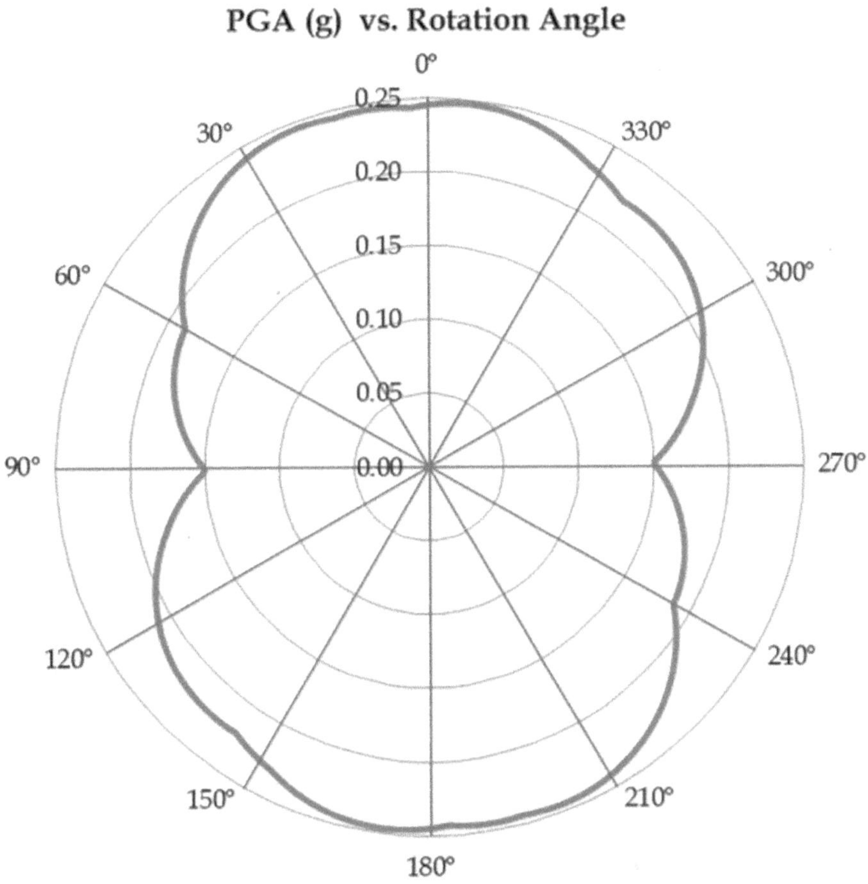

FIGURE 7.26 Ground motion directionality – PGA.

The envelope of all those rotated spectra is *RotD100*. The median of all those spectra is *RotD50*. The minimum of those spectra is *RotD00*:

$$a(\theta, t) = a_X(t)\cos\theta + a_Y(t)\sin\theta \tag{7.17}$$

ASCE 7-22, in Article 21.2, further specifies that the ratio of maximum-direction spectra to geometric mean spectra be determined as follows, when conversion between the two bases is required:

- for periods less than or equal to 0.20 seconds, $PSA_{RotD100}/PSA_{GeoMean} = 1.20$
- at a period of 1.0 second, $PSA_{RotD100}/PSA_{GeoMean} = 1.25$
- for periods greater than or equal to 10.0 seconds, $PSA_{RotD100}/PSA_{GeoMean} = 1.3$
- for period values between those specified above, a linear interpolation is to be used.

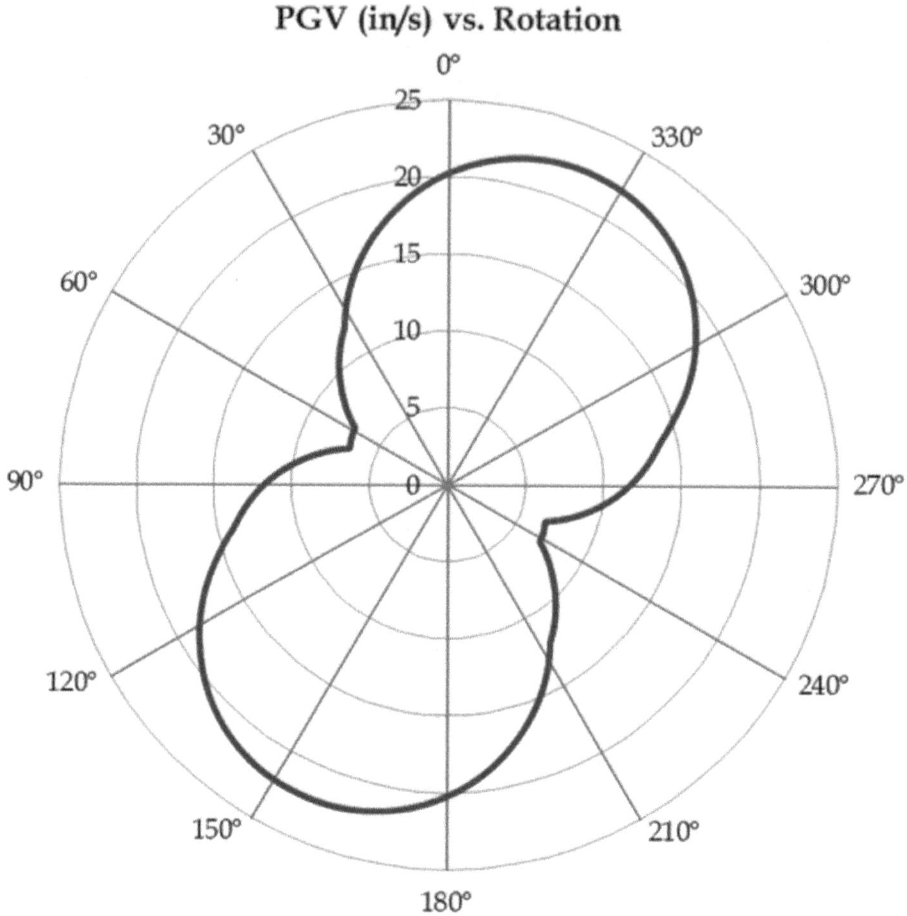

FIGURE 7.27 Ground motion directionality – PGV.

The DOS-based Time Series Processing Programs, TSPP (Boore, 2020), are useful in computation of response spectra for rotated ground motion record pairs. For a given recorded record pair, a new record pair could be computed and represent those that would have been obtained had the instrumentation been oriented differently. The effect of rotation upon the geometric mean response at a period of 1.0 seconds for two record pairs, PEER RSN 0900 from the Landers earthquake and PEER RSN 1147 from the Kocaeli earthquake, is illustrated in Figure 7.29. So, as evident in the Figure, the effect may be relatively mild (Kocaeli RSN 1147) or relatively strong (Landers RSN 0900). Both records were obtained from the NGA-West2 ground motion database (see Section 7.4.1).

Also available in TSPP is the capability to compute various ground motion parameters as well as pseudo-acceleration-spectra, such as the geometric mean (*GeoMean*), *RotD50*, *RotD100*, and others. The *nga2psa_rot_gmrot* utility provides direct

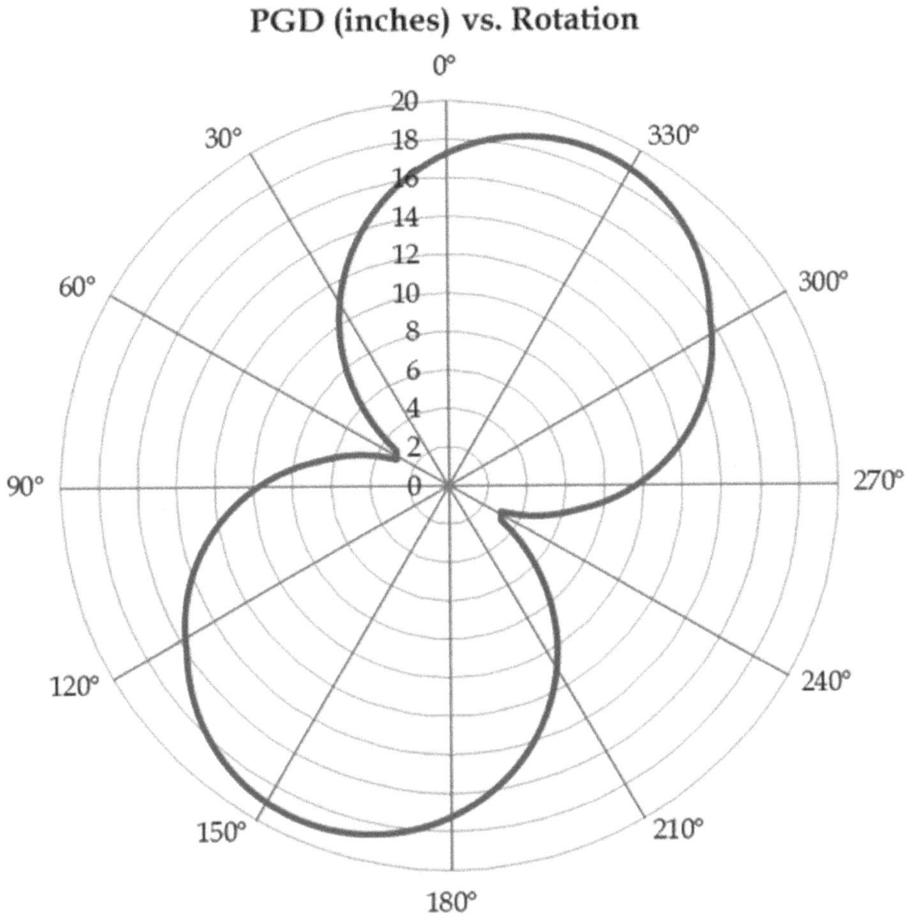

FIGURE 7.28 Ground motion directionality – PGD.

calculation of spectra and ground motion parameters from NGA-formatted ground motion record pairs. It is instructive to observe *GeoMean*/*RotD50* and *RotD100*/*RotD50* spectral ratios for various ground motion suites. Figures 7.30 and 7.31 depict these ratios and their variability over a wide period range for a large number of records and qualitatively confirm findings in the literature (Boore & Kishida, 2017) that (1) *RotD50* and *GeoMean* are similar, (2) *RotD100* to *GeoMean* ratios are in the approximate range of 1.2–1.3 for periods of 1 second and more, and (3) *RotD100* to *GeoMean* ratios are about 1.1–1.2 for short periods.

Figure 7.32 shows the *RotD100*-to-*GeoMean* ratio averaged for 45,749 records from the NGA Subduction flatfile. Also shown in Figure 7.32 is the *RotD100*-to-*GeoMean* ratio specified in ASCE 7-22, Section 21.2. The Figure suggests that the ratio may be somewhat higher for subduction zone earthquakes compared to earthquakes from shallow crustal faults.

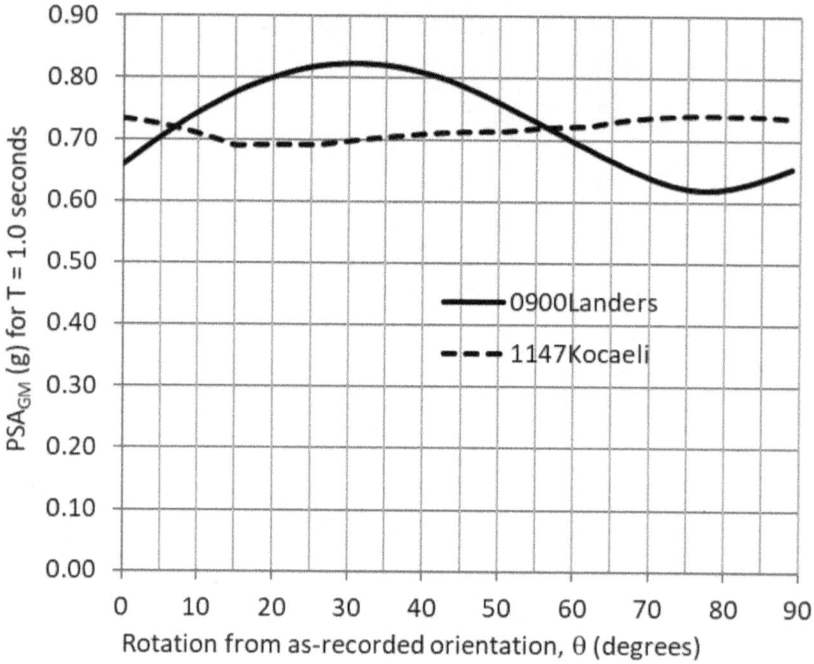

FIGURE 7.29 Effect of rotation on geometric mean response.

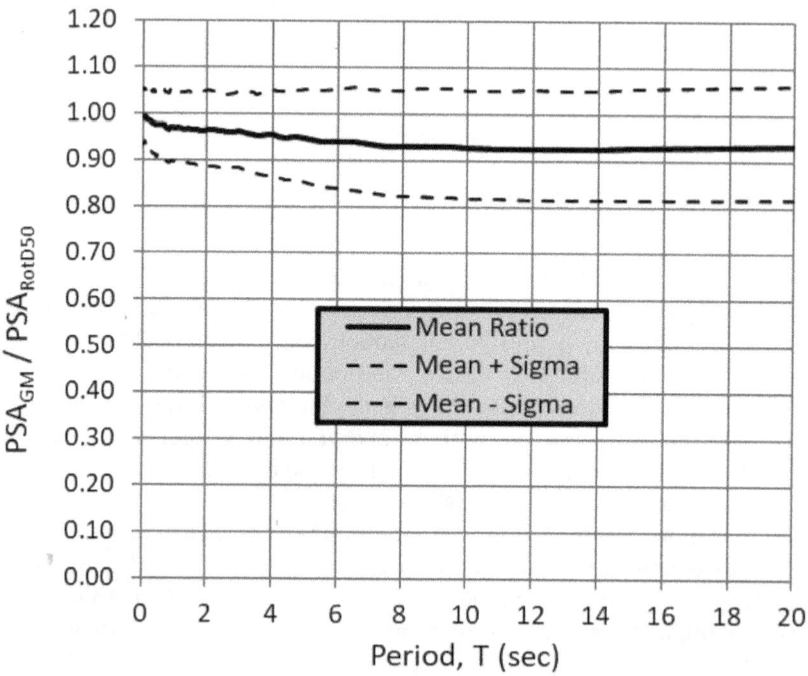

FIGURE 7.30 *GeoMean* to *RotD50* spectra ratio.

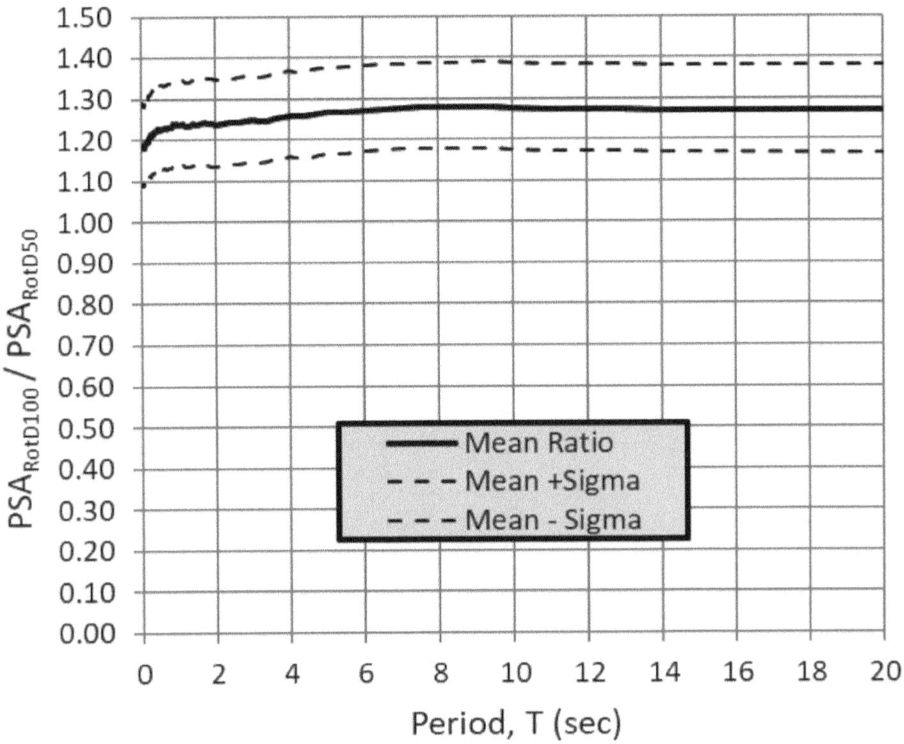

FIGURE 7.31 *RotD100* to *RotD50* spectral ratio.

7.7.2 Elastic Response Spectra

A response spectrum is that which may be inferred from the name – a spectrum of responses. For linear analysis-based response spectra, the only required inputs are a ground motion accelerogram and the system damping, expressed as a fraction of critical damping. Applied to earthquakes, ground motion response spectra from linear analysis of a single-degree-of-freedom (SDOF) oscillator include:

- spectral displacement, SD
- spectral velocity, SV
- spectral acceleration, SA
- pseudo-spectral velocity, PSV
- pseudo-spectral acceleration, PSA

Consider, first, a single-degree-of-freedom (SDOF) system subjected to a ground motion which varies with time, as shown in Figure 7.33. The equation of motion for a linear system is given by Equation 7.18 and must be solved numerically since the loading history defined by an earthquake ground acceleration is somewhat random, rather than being defined by a closed form function. Structural dynamics theory reveals that the structural response for a given ground acceleration history, x, depends

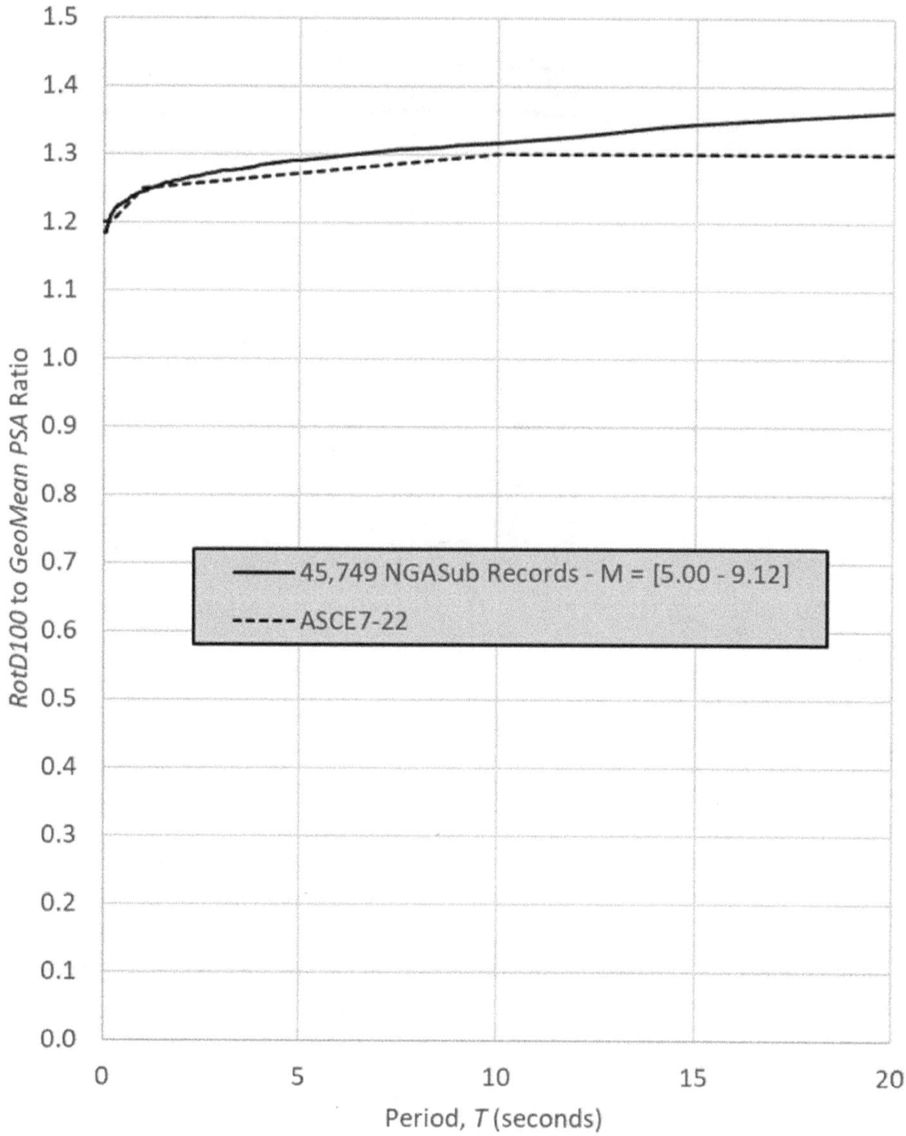

FIGURE 7.32 *RotD100*-to-*GeoMean PSA* for subduction records.

only upon two parameters, the natural period of the system (T) and the fraction (ξ) of critical damping (ξ_{cr}). Damping is energy dissipation which makes a vibrating system eventually stop vibrating:

$$\ddot{x}(t) + 2\xi\omega_n\dot{x}(t) + \omega_n^2 x(t) = -\ddot{x}_g(t) \tag{7.18}$$

$$\omega_n = \sqrt{\frac{k}{m}}, \quad \text{natural frequency in rad/s} \tag{7.19}$$

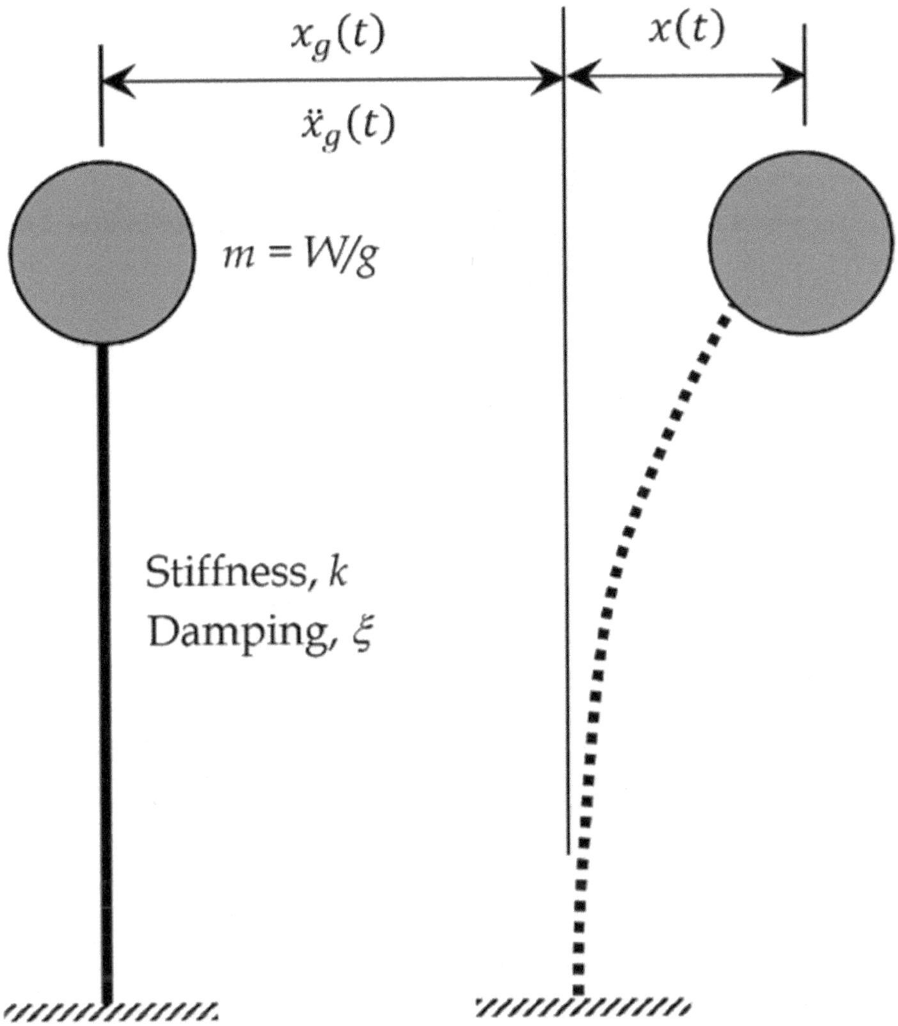

FIGURE 7.33 SDOF system properties.

$$f_n = \frac{\omega_n}{2\pi}, \quad \text{natural frequency in cycles/ s (Hz)} \tag{7.20}$$

$$T = \frac{1}{f_n}, \quad \text{natural period in seconds} \tag{7.21}$$

$$\xi_{cr} = 2m\omega_n = 2\sqrt{\text{km}} \tag{7.22}$$

Figure 7.34 depicts the ground acceleration history for the East-West component recorded at the "LLO" station during the 2010 M_W 8.8 Maule, Chile earthquake.

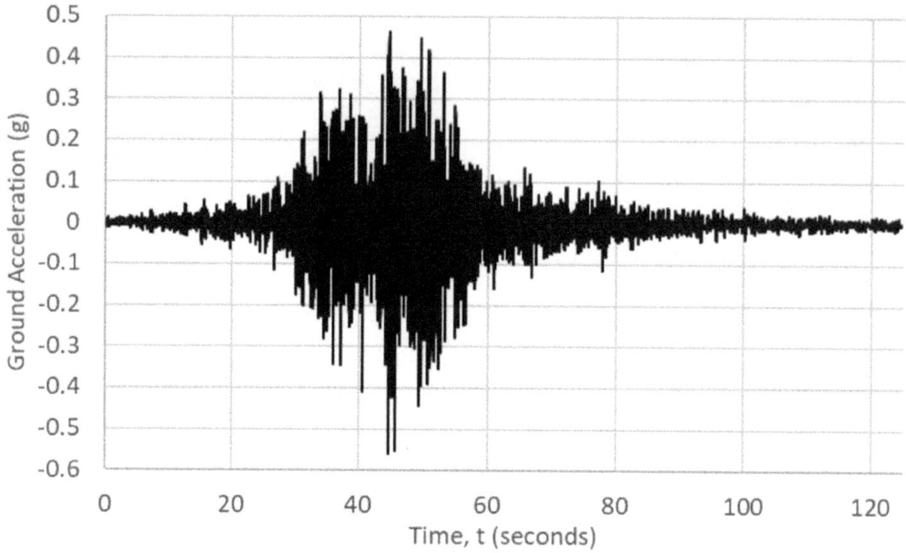

FIGURE 7.34 Maule, Chile LLO east-west ground motion accelerogram.

This represents the right side of Equation 7.18. SeismoSignal or PRISM (or other software) may be used to obtain the response history of a linear SDOF system having a natural period of 1.0 second and 5% damping. Using PRISM for the solution produces the following maxima:

$$\ddot{x}_{\max} = 0.675 \text{ g}$$

$$\dot{x}_{\max} = 108 \text{ cm/s}$$

$$x_{\max} = 16.7 \text{ cm}$$

These maxima represent one point (at a period of 1 second) on the acceleration, velocity, and displacement spectra plots, respectively.

The process could be repeated for SDOF systems having various natural periods and the maxima of acceleration, velocity, and displacement recorded. A plot of each parameter versus natural period results in the response spectra for that ground motion. Figure 7.35 shows the spectral acceleration and the pseudo-spectral acceleration for the ground motion represented in Figure 7.34 from the Maule, Chile earthquake. Spectral velocity and pseudo-spectral velocity are shown in Figure 7.36 and spectral displacement is shown in Figure 7.37. Each of the response spectra have been computed out to a period of 6 seconds.

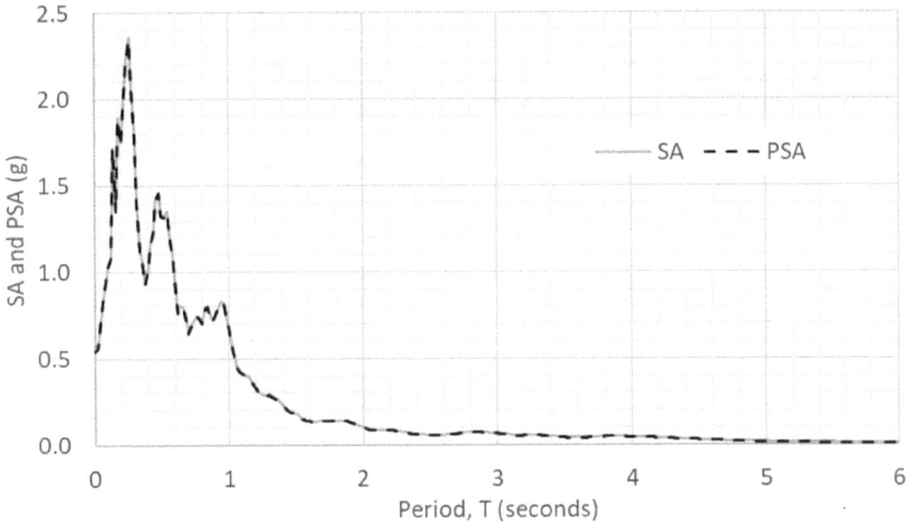

FIGURE 7.35 Spectral acceleration and pseudo-spectral acceleration.

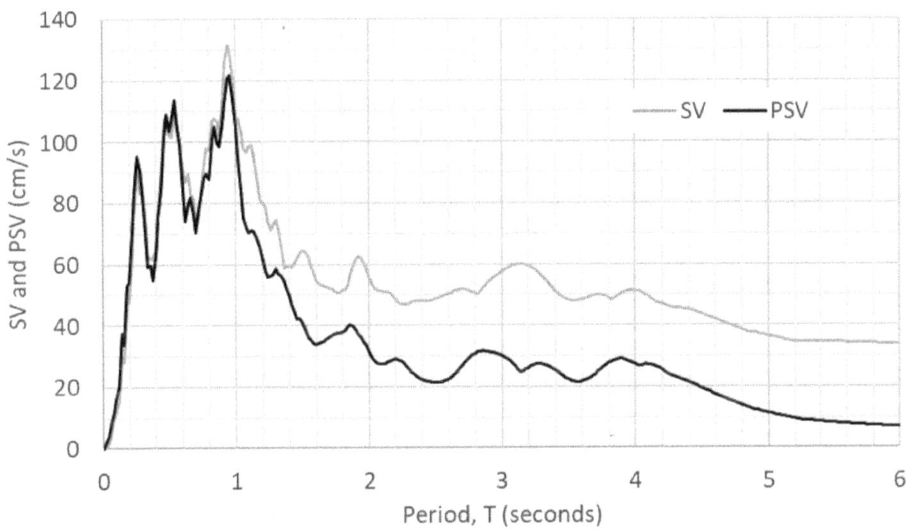

FIGURE 7.36 Spectral velocity and pseudo-spectral velocity.

Spectral acceleration (SA) is the absolute acceleration of an oscillator obtained from a direct solution of the equations of motion.

Spectral velocity (SV) is the relative (to the ground) velocity of an oscillator obtained from a direct solution of the equations of motion.

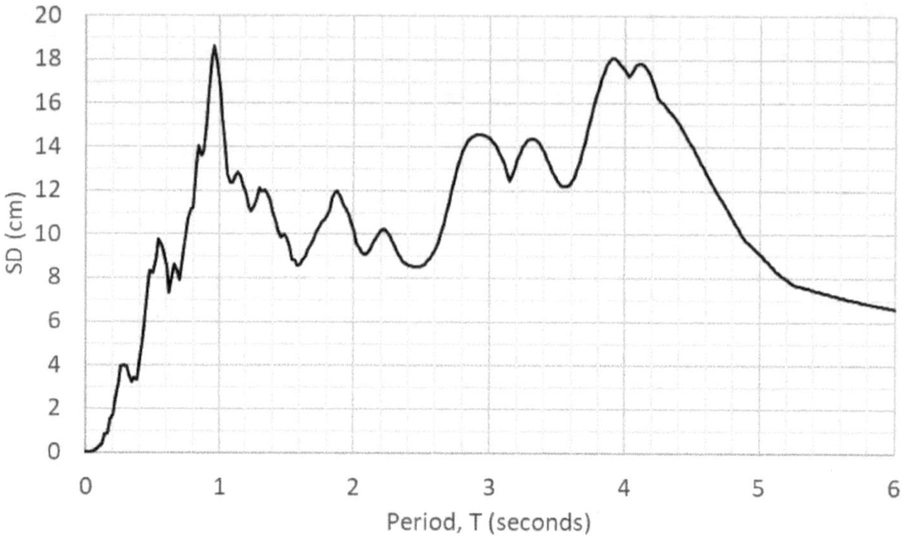

FIGURE 7.37 Spectral displacement.

Spectral displacement (SD) is the relative (to the ground) displacement of an oscillator obtained from a direct solution of the equations of motion. The relative displacement is the displacement of structural interest in building and bridge (or any other structure) design.

Pseudo-spectral-velocity (PSV) and pseudo-spectral-acceleration (PSA) are values calculated from the spectral displacement (SD) assuming a harmonic relationship.

$$PSV = SD \cdot \frac{2\pi}{T} \tag{7.23}$$

$$PSA = SD \cdot \left(\frac{2\pi}{T}\right)^2 \tag{7.24}$$

PSV is often very close to SV over a wide range of periods. Beyond a period of about 1 second, this is not the case in Figure 7.36 however.

PSA is often very close to SA over a wide range of periods. For the full range of periods up to 6 seconds, the two are virtually indistinguishable in Figure 7.35. This will not always hold true.

Given that modern seismic design principles are focused on the estimation of displacements, it makes sense to adopt the pseudo-spectral acceleration (as opposed to the spectral acceleration) as the design basis. Pseudo-spectral acceleration may be used in Equation 7.24 to recover spectral displacement, one of the most critically important parameters in modern seismic design. The same may not be said for spectral acceleration – spectral displacement is not necessarily recovered by using the harmonic relationship of Equation 7.24 with SA rather than PSA.

7.7.3 Inelastic Response Spectra

For single-degree-of-freedom systems which have a defined yield force, rather than infinite strength, it is possible to construct inelastic acceleration, velocity, and displacement spectra for ground motion records. One of the most common forms for inelastic spectra is the constant ductility response spectrum. A desired level of displacement ductility is selected, based on design requirements, and iterative algorithms permit the computation of the spectral parameters at each period corresponding to that ductility.

For linear analysis-based response spectra, the only required inputs are the ground motion accelerogram and the system damping. For constant ductility inelastic spectra from ground motion accelerograms by response history analysis, a hysteretic model and a specified ductility are required as well. Ductility, in this context, refers to displacement ductility. There are other ductility definitions – rotation, curvature, or strain, for example.

Displacement ductility, μ, is defined as maximum displacement divided by yield displacement for a bilinear system. Figure 7.38 graphically depicts the maximum displacement, Δ_{INEL}, and the yield displacement, Δ_y for a bilinear hysteretic model. The

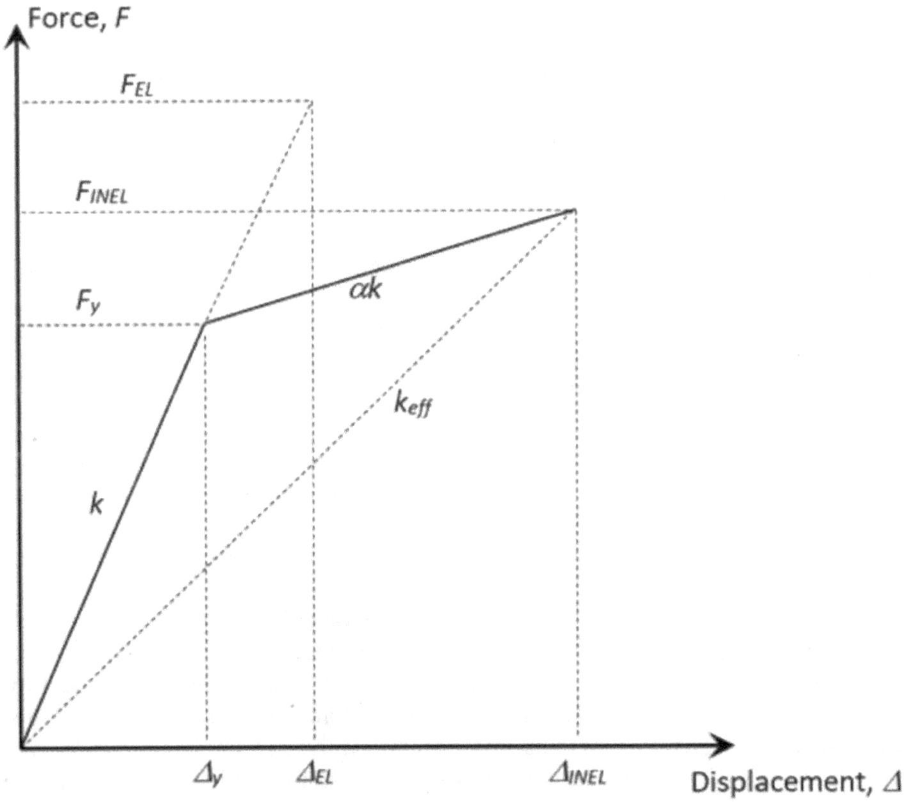

FIGURE 7.38 Bilinear force-displacement parameters.

ratio of maximum displacement, Δ_{INEL}, to that displacement which would have been experienced if nothing yielded, Δ_{EL}, is called a displacement amplification factor, C_μ. Figure 7.38 represents a simple, bilinear hysteretic model. More complex models are available in modern software. Nonetheless, the bilinear model is frequently adopted in nonlinear analytical procedures.

Research has suggested that the incorporation of the traditional $\xi_o = 5\%$ damping (0.05 times critical damping), while appropriate for elastic response spectra generation from accelerograms for some systems, may introduce spurious, artificial damping when used in nonlinear response history analysis. One suggested method for mitigating this effect is to reduce the initial, elastic damping portion of the total damping. Total damping in a nonlinear response history analysis consists of the initial, elastic component of damping, ξ_{el}, and the hysteretic component which arises from inelastic response. The suggested reduction in the initial, elastic component is given by Equation 7.25 (Priestley & Grant, 2005). The post-yield stiffness ratio is α (see Figure 7.38) and the displacement ductility is μ in Equation 7.25:

$$\xi_{el} = \frac{\xi_o\left[1 - 0.1(\mu - 1)(1 - \alpha)\right]}{\sqrt{\dfrac{\mu}{1 + \alpha\mu - \alpha}}} \tag{7.25}$$

Figure 7.39 shows the variation of the suggested damping as a function of displacement ductility for four different post-yield stiffness values. Observe, for example, that the recommended initial elastic component of damping is about 20% of the traditional value for an elastic-perfectly-plastic (EPP, $\alpha = 0$) system at a ductility of 6.

Figure 7.40 is one example of a constant ductility inelastic displacement spectrum generated form a ground motion accelerogram, namely the East-West component of the ground motion recorded at the LLO station during the 2010 Maule M_W 8.8 earthquake. The plot is for a ductility equal to 6 with 1% initial, elastic damping. A bilinear hysteretic model with zero post-yield stiffness (elastic-perfectly-plastic) was adopted.

Observe from Figure 7.40 that the inelastic displacement may be equal to, less than, or greater than the displacement corresponding to the elastic case for an individual ground motion. Figure 7.41 depicts the ratio of inelastic to elastic displacement spectra for the ground motion being considered. For very short periods, this ratio approaches the displacement ductility.

It is possible to generate an estimated constant ductility inelastic response spectrum directly from an elastic design response spectrum rather than from a suite of ground motion accelerograms. Inelastic behavior effectively adds damping to the system. An increase in effective damping produces a decrease in response. This type of analysis requires both a model for effective damping and a model for response modification due to effective damping. The analysis is referred to as a substitute structure method (SSM) analysis since an equivalent single-degree-of-freedom system with effective damping and stiffness is employed in a simple response spectrum analysis.

The generation of inelastic spectra from a suite of accelerograms is accomplished through nonlinear response history analysis (NLRHA) across a range of periods.

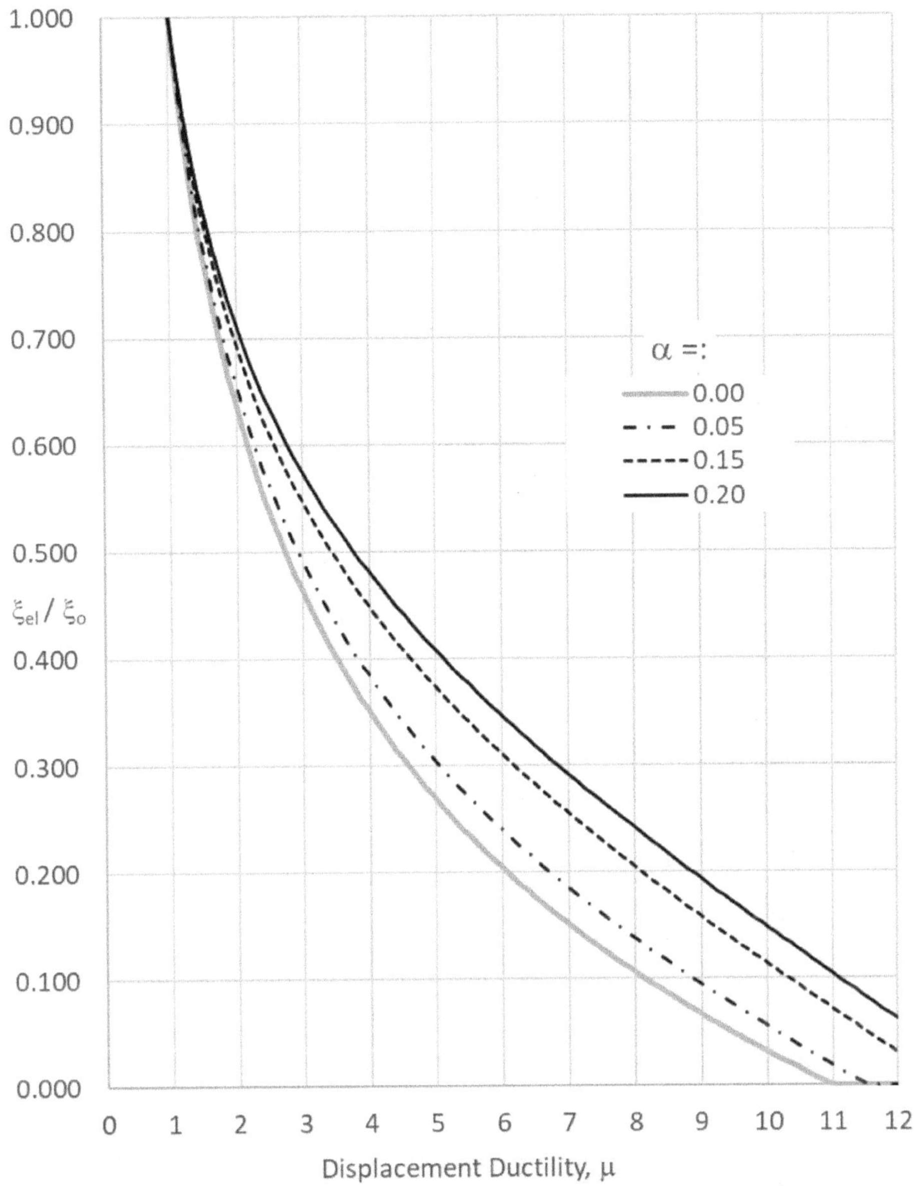

FIGURE 7.39 Suggested NLRHA initial elastic damping ratios.

The NLRHA loading consists of a suite of properly selected and modified ground motion record pairs scaled or matched to the target response spectrum. The SSM analysis does not require the selection and modification of a suite of accelerograms.

One of the models proposed for effective damping in this type of SSM analysis is given in Equation 7.26 (Priestley et al., 2007; Priestley & Grant, 2005).

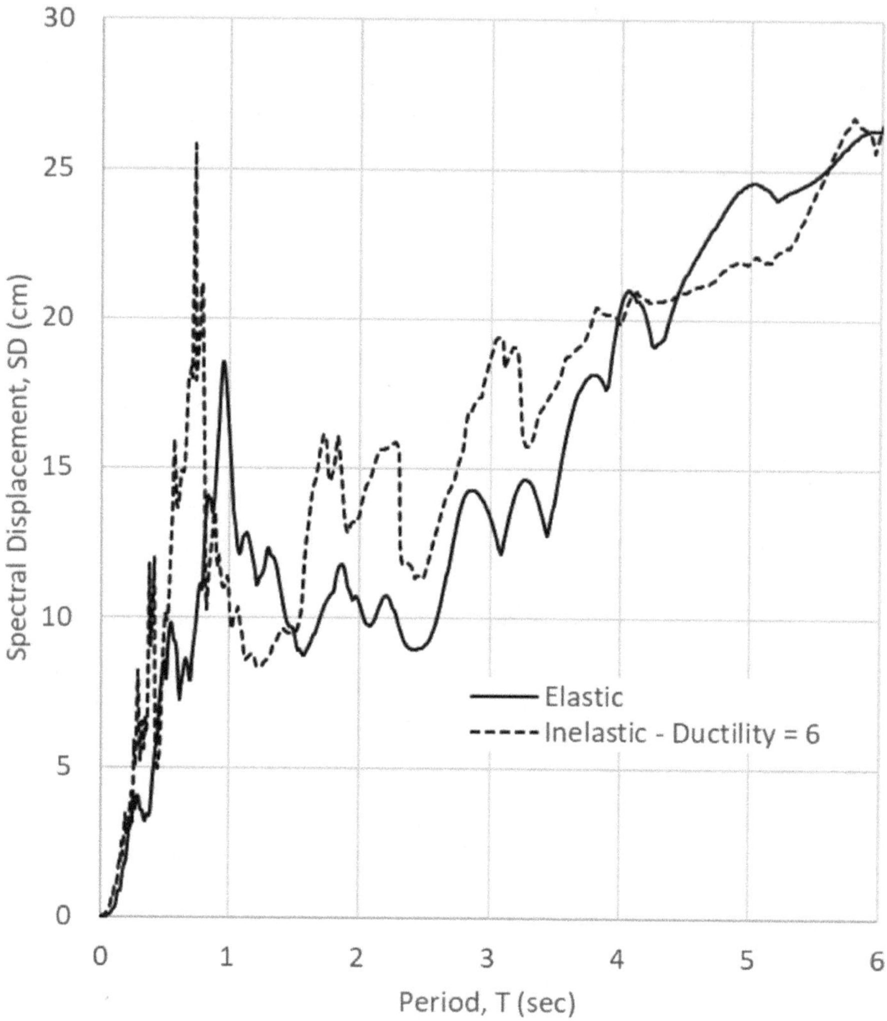

FIGURE 7.40 Inelastic displacement spectrum for the LLO east-west record.

The coefficients required for Equation 7.26 depend on the hysteretic model adopted and are summarized in Table 7.17. The effective period, T_{EFF}, is based on a secant stiffness and is given by Equation 7.27, where T_i is the initial-stiffness-based natural period:

$$\xi_{\text{eff}} = \xi_o + \xi_{\text{hys}} = \mu^{\lambda}\xi_{\text{el}} + a\left(1 - \frac{1}{\mu^b}\right)\left(1 + \frac{1}{(T_{\text{EFF}} + c)^d}\right) \quad (7.26)$$

$$T_{\text{EFF}} = T_i\sqrt{\frac{\mu}{1 + \alpha\mu - \alpha}} = \gamma_T T_i \quad (7.27)$$

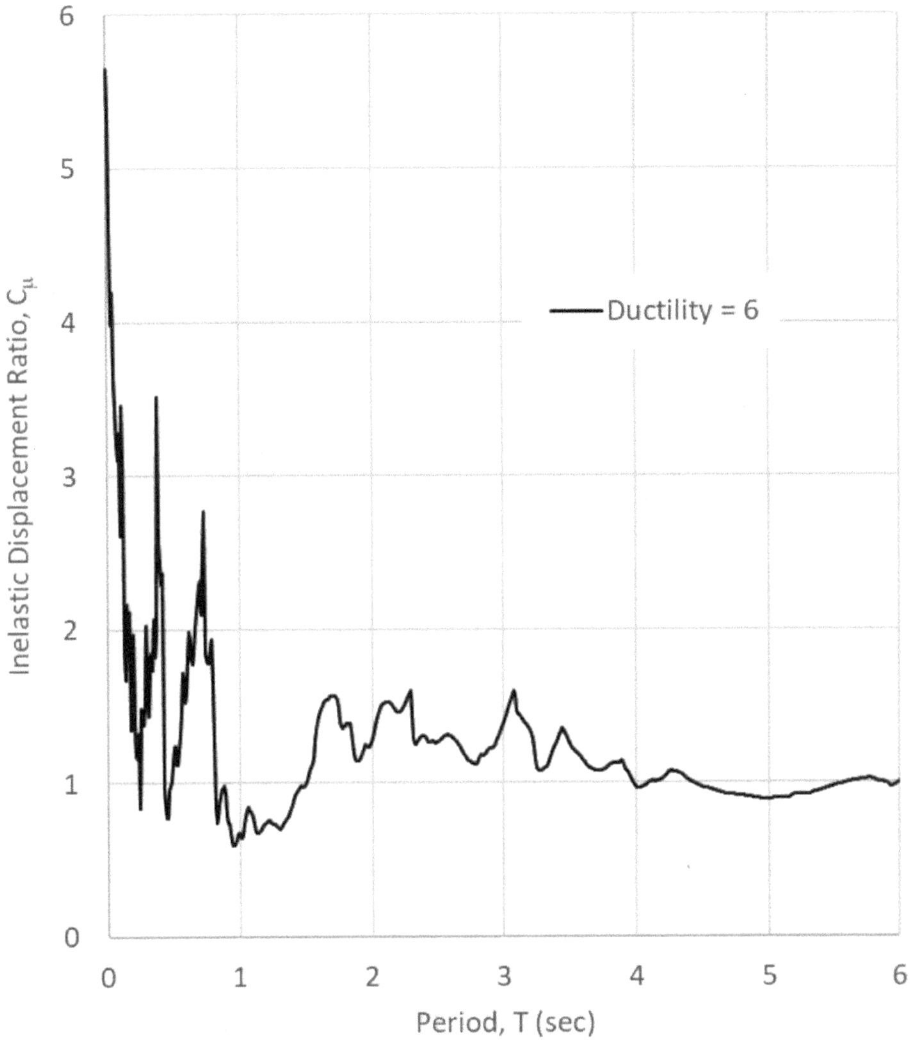

FIGURE 7.41 Inelastic displacement ratio for LLO – ductility = 6.

- EPP: Elastic-Perfectly-Plastic (bilinear with $\alpha = 0.00$) – recommended use is friction sliding isolation systems
- BI: Bilinear with $\alpha = 0.20$
- TT: Takeda Thin – recommended use is with bridges
- TF: Takeda Fat – recommended use is with concrete frame buildings
- FS: Flag-Shaped – recommended use is with hybrid prestressed concrete frames
- RO: Ramberg-Osgood – recommended use is with steel frame buildings

TABLE 7.17
Coefficients for Priestley & Grant's Effective Damping Model

Model	*a*	*b*	*c*	*d*	*λ*
EPP	0.224	0.336	−0.002	0.250	−0.341
BI	0.262	0.655	0.813	4.890	−0.808
TT	0.215	0.642	0.824	6.444	−0.378
TF	0.305	0.492	0.790	4.463	−0.313
FS	0.251	0.148	3.015	0.511	−0.430
RO	0.289	0.622	0.856	6.460	−0.617

Other models include a theoretical formulation for bilinear systems, given by Equation 7.28 as well as simpler forms, compared to Grant's model, given by Equations 7.29 through 7.32:

$$\xi_{\text{eff}} = \xi_o + \frac{2(\mu-1)(1-\alpha)}{\pi\mu(1+\alpha\mu-\alpha)} \tag{7.28}$$

$$\xi_{\text{eff}} = 5 + 44.4\frac{\mu-1}{\pi\mu} \text{ (concrete bridge columns)} \tag{7.29}$$

$$\xi_{\text{eff}} = 5 + 57.7\frac{\mu-1}{\pi\mu}, \quad \text{steel framed buildings} \tag{7.30}$$

$$\xi_{\text{eff}} = 5 + 67.0\frac{\mu-1}{\pi\mu}, \quad \text{EPP} \tag{7.31}$$

$$\xi_{\text{eff}} = 5 + \frac{120}{\pi}\left(1 - \frac{1}{\sqrt{\mu}}\right), \quad \text{concrete framed buildings} \tag{7.32}$$

Various models for response modification due to effective damping have been proposed as well (Priestley et al., 2007). To account for added effective damping, a response modifier, R_ξ, is needed. Equations 7.33 through 7.37 provide five such models. Equation 7.37 is a unique model (Stafford et al., 2008) in that the significant duration of ground shaking, D_{5-95}, appears explicitly in the model. Refer to Section 7.5.3 for the definition of the significant duration, D_{5-95}:

$$B_L = \frac{1}{R_\xi} = \left(\frac{\xi_{\text{eff}}}{0.05}\right)^{0.30} \le 1.70, \quad \text{AASHTO (Isolated Bridges)} \tag{7.33}$$

$$R_\xi = \left(\frac{0.05}{\xi_{\text{eff}}}\right)^{0.40}, \quad \text{AASHTO (Conventional Bridges)} \tag{7.34}$$

$$R_\xi = \left(\frac{0.10}{0.05+\xi}\right)^{0.50}, \quad \text{Eurocode} \tag{7.35}$$

$$R_\xi = \left(\frac{0.07}{0.02+\xi}\right)^{0.25}, \quad \text{Velocity pulse conditions} \tag{7.36}$$

$$R_\xi = 1 - \frac{-0.631 + 0.421\ln(100\xi) - 0.015\ln(100\xi)^2}{1+\exp\left\{\frac{-\left[\ln(D_{5-95\%}) - 2.047\right]}{0.930}\right\}} \tag{7.37}$$

While the elastic response spectrum for a given ground acceleration history depends only upon natural period (T) and damping fraction (ξ), the generation of an inelastic response spectrum directly from the elastic response spectrum depends additionally upon the choice of hysteretic model, the model selected for effective damping, and the model selected for response modification.

The SSM could not necessarily be expected to consistently and precisely estimate inelastic displacements over a wide period range for a single ground motion record. However, for suites of ground motions scaled or spectrally matched to a design response spectrum, the SSM seems to be a promising tool for generating inelastic displacement spectra directly from elastic displacement spectra. The example problem discussed next will illustrate this.

7.7.3.1 Example Problem: Ground Motion Selection and Modification for NLRHA and SSM

Suppose a two-point design response spectrum has been given from the control points, $S_{DS}=0.643$ g, $S_{D1}=0.473$ g, and the corner period $T_L=12$ seconds (Figure 7.21). Ground motion selection and modification is a developing science, and there are alternatives for target response spectra. For the examples presented in this book, the code-based uniform hazard-based response spectra are adopted as targets for ground motion modification.

A relatively wide range of magnitudes has been used for this example. Disaggregation of the seismic hazard at a site is useful in establishing the magnitude and distance parameters appropriate for a given site. Disaggregation should always be a part of ground motion selection.

A period range of interest equal to [0.20–3.00] seconds has been determined for the project. A suite of 30 ground motion record pairs has been selected and scaled to the design response spectrum, which is geometric mean-based. So, the geometric mean of each record pair has been scaled to the target response spectrum. Figure 7.42 shows the suite geometric mean and the target response spectra up to a period of 6 seconds after conversion from a PSA-based to a SD-based spectrum. The suite matches the target closely in the period range of interest.

Suppose further that an inelastic displacement spectrum is needed for the project and that the system modeled is a bilinear hysteretic behavior with post-yield stiffness equal to 0.005. The target displacement ductility for the structural design is $\mu=6.0$.

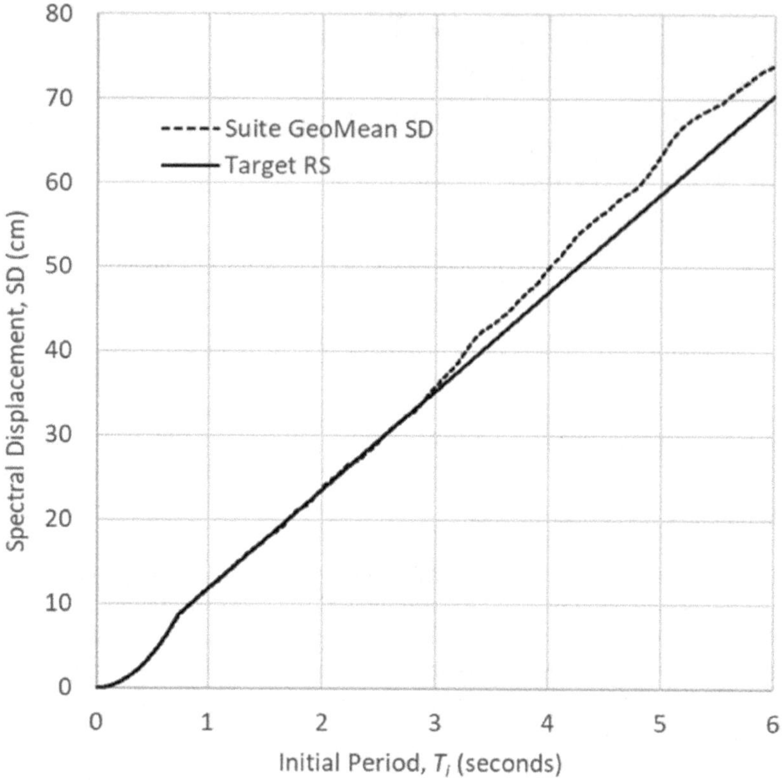

FIGURE 7.42 Suite mean and target response spectra.

Figure 7.43 shows the same target response spectrum. Also shown are the inelastic displacement spectrum based on the average response for the 30-record suite and the inelastic displacement spectrum based on the equivalent linear analysis using Equation 7.31 for effective damping ($\alpha = 0.005$ is approximated by EPP hysteretic behavior) and Equation 7.34 for response modification due to effective damping. For the inelastic response history analysis using the 30 record pairs, an initial damping of 1% was used, as obtained from Equation 7.25 with $\alpha = 0.005$ and $\mu = 6$:

$$\xi_{\text{eff}} = 5 + 67.0\frac{6-1}{6\pi} = 22.8\% \Rightarrow \xi_{\text{eff}} = 0.228$$

$$R_{\xi} = \left(\frac{0.05}{0.228}\right)^{0.40} = 0.545$$

$$T_{\text{EFF}} = T_i\sqrt{\frac{6}{1+0.005\times 6-0.005}} = 2.419T_i$$

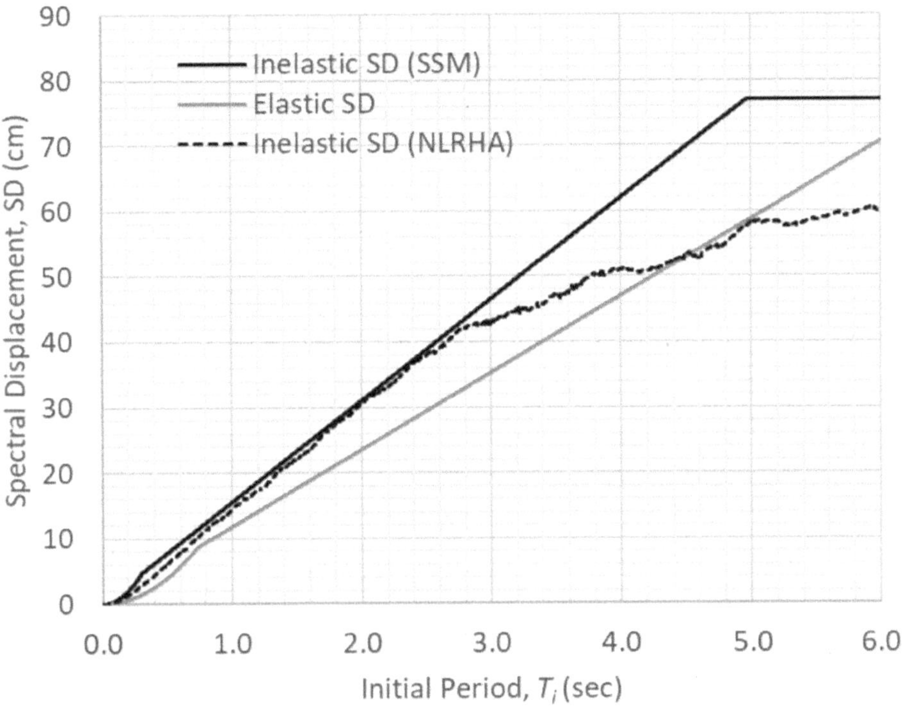

FIGURE 7.43 NLRHA versus SSM inelastic spectra.

Figure 7.43 shows the comparison between nonlinear response history analysis results for the 30-record-pair suite versus the substitute structure method solution using secant stiffness, effective damping, and response modification from effective damping. The SSM analysis provides an effective solution within the period range of interest. The largest deviation occurs at the high end of the period range (3.0 seconds) where the SSM solution exceeds the NLRHA solution by 9% (46.5 cm versus 42.7 cm).

Notice that at an initial period of 4.96 seconds, the inelastic spectral displacement plot from the SSM analysis becomes horizontal. This is because the effective period at this point is $4.96 \times 2.419 = 12$ seconds, the corner period for the design response spectrum defining the onset of constant spectral displacement.

The key to the SSM is selection of the appropriate effective damping model along with the appropriate response modification model.

SSM analysis may also be used to establish estimates for the displacement amplification due to inelastic behavior, C_μ. Figure 7.44 is the plot of displacement amplification from inelasticity, C_μ, versus initial period, T_i, for the SSM results depicted in Figure 7.43.

In effect, the SSM analysis permits the inclusion of nonlinear effects through the use of a linear response spectrum analysis.

Comparing the natural-logarithm-based standard deviation for the elastic spectra and the inelastic spectra, there is more variability in the inelastic response than in the elastic response for this example.

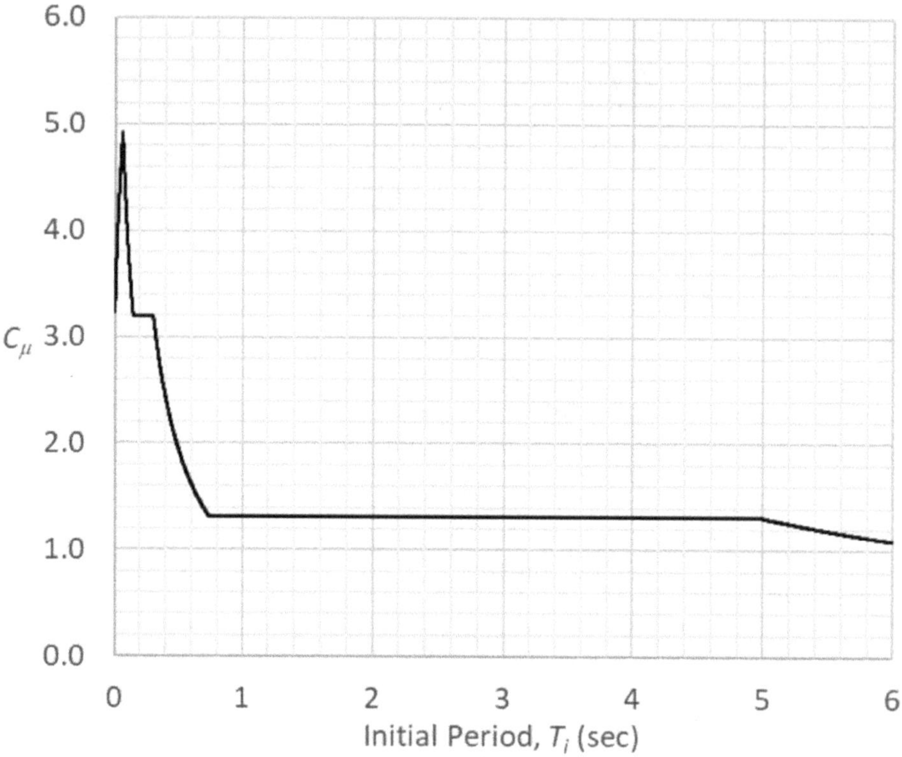

FIGURE 7.44 Inelastic displacement amplification factor, C_μ.

$\sigma_{\text{lnSD-EL}} = 0.601$ average, 0.661 maximum, 0.543 minimum with the period range of interest

$\sigma_{\text{lnSD-INEL}} = 0.624$ average, 0.754 maximum, 0.538 minimum with the period range of interest

The 30 record pairs, along with pertinent metadata, are summarized in Table 7.18. RSN is the PEER Record Sequence Number. The records were obtained from the PEER Ground Motion Database (Pacific Earthquake Engineering Research Center, 2013). The target response spectrum along with a target natural-logarithm-based standard deviation of 0.6 across all periods in the period range of interest, were used to scale the records.

From the SSM analysis with non-period-dependent effective damping, it can be shown that the inelastic displacement amplification, C_μ, is given by Equation 7.38 for all periods between T_S ($= S_{D1}/S_{DS}$) and T_L/γ_T. The peak on the C_μ plot approaches μ. The maximum inelastic spectral displacement is given by Equation 7.39. See Equation 7.27 for γ_T.

$$C_\mu = \gamma_T R_\xi \quad \text{for} \quad T_S \le T_i \le {}^{T_L}\!/_{\gamma_T} \tag{7.38}$$

$$\left(\text{SD}_{\text{inel}}\right)_{\max} = \frac{S_{D1} \cdot g \cdot T_L \cdot R_\xi}{4\pi^2} \ @\ T_i = {}^{T_L}\!/_{\gamma_T} \tag{7.39}$$

TABLE 7.18
Thirty-Record-Pair Ground Motion Suite

RSN	Earthquake	M_W	Scale Factor	$D_{5\text{–}95}$, seconds
169	1979 Imperial Valley		2.5615	50.7
1196	1999 Chi-Chi	7.62	3.9279	44.9
1207	1999 Chi-Chi	7.62	3.3002	43.3
1226	1999 Chi-Chi	7.62	3.1549	46.9
1331	1999 Chi-Chi	7.62	3.2963	41.5
1414	1999 Chi-Chi	7.62	2.8423	42.1
1188	1999 Chi-Chi	7.62	2.9422	71.0
1192	1999 Chi-Chi	7.62	3.8728	51.7
1213	1999 Chi-Chi	7.62	3.1529	59.0
5117	2007 Chuetsu-oki	6.60	3.4508	44.1
5260	2007 Chuetsu-oki	6.60	3.2883	46.2
5856	2010 El Mayor	7.20	4.3847	68.8
5823	2010 El Mayor	7.20	2.7506	51.2
5989	2010 El Mayor	7.20	3.1530	66.5
5880	2010 El Mayor	7.20	4.6932	48.4
5988	2010 El Mayor	7.20	2.7373	56.3
5860	2010 El Mayor	7.20	3.7514	49.9
8492	2010 El Mayor	7.20	3.2391	50.7
5846	2010 El Mayor	7.20	4.2014	45.5
5859	2010 El Mayor	7.20	3.0772	65.3
1823	1999 Hector Mine	7.10	3.8608	57.3
3789	1999 Hector Mine	7.10	4.9522	43.4
3811	1999 Hector Mine	7.10	4.1373	41.8
5814	2008 Iwate	6.90	3.7222	41.0
5628	2008 Iwate	6.90	6.3045	47.7
5665	2008 Iwate	6.90	2.8040	44.6
5683	2008 Iwate	6.90	4.5949	42.8
5746	2008 Iwate	6.90	4.4560	56.1
1100	1995 Kobe	7.00	3.0788	53.1
4151	2004 Niigata	6.60	3.7778	56.9

7.7.4 Design Response Spectra Data and Tools

When earthquake response spectra are needed for design, distinction needs to be made between uniform hazard design response spectra (DRS) and risk-targeted design response spectra. Historically, a uniform hazard-based DRS has been adopted, and still is adopted in some design specifications (ASCE 43-19, for example). More recent DRS are often risk-targeted in nature (ASCE 7-22, for example).

The probability of ground motion exceedance is constant across all periods for a uniform hazard spectrum. The risk of collapse is uniform across all periods for a risk-targeted response spectrum.

DRS are smoothed curves developed using seismological theory and known fault characteristics for a given site. Historic seismic data are combined with fault propagation theory to develop ground motion models. These ground motion models (also known as ground motion prediction equations or attenuation models) are used to perform a seismic hazard analysis for a given site.

The United States Geological Survey (USGS) is intimately involved in the development of ground motion tools. It is the responsibility of the engineer on a given project to generate the appropriate DRS for design.

Some of the tools available for the generation of ground motion design response spectra and related data are summarized here.

a. The ASCE 7 Hazard Tool (https://asce7hazardtool.online/) reports DRS for buildings using criteria from ASCE 7-10, ASCE 7-16, or ASCE 7-22. Ground motion DRS may also be obtained for seismic retrofitting of buildings in accordance with either ASCE 41-17 or ASCE 41-23.
b. The Structural Load Data Tool for UFC 3-301-01 (https://www.wbdg.org/tools/sldrt) provides DRS data for US military base sites and international locations. Available standards include ASCE 7-10, ASCE 7-16, ASCE 41-17, and IBC 2015.UFC 3-301-01 (USDOD, 2023) contains structural design criteria for Department of Defense structures.
c. The 2020 National Building Code of Canada Seismic Hazard Tool (https://earthquakescanada.nrcan.gc.ca/hazard-alea/interpolat/nbc2020-cnb2020-en.php) provides DRS values for buildings in Canada in accord with the 2020 National Building Code of Canada.
d. The USGS-based (https://earthquake.usgs.gov/ws/designmaps/aashto-2023/) AASHTO-2023 online tool is an excellent starting point to gather DRS data for bridges based on 2023 AASHTO design specifications.
e. Other tools available from the USGS (https://earthquake.usgs.gov/nshmp/) include various applications for disaggregation of seismic hazard at a given site. Disaggregation is the process of identifying the underlying earthquake magnitude and source-to-site metrics for a given location. Disaggregation is necessary when ground motion selection and modification procedures are required to develop a suite of accelerogram pairs to be used in response history analysis of structures. In such cases, the DRS is needed as a target for the elastic response spectra of accelerogram pairs through various modification procedures (amplitude scaling of the accelerograms, spectral matching in the time domain through the addition of wavelets, or spectral matching in the frequency domain through manipulation of the Fourier response spectrum of the accelerograms). When design is accomplished through dynamic modal superposition response spectrum analysis, only the DRS is needed, and disaggregation is typically unnecessary.

The availability of ground motion tools for design has grown tremendously over the years and continues to do so. Currently available tools evolve over time and new tools become available frequently. It is important for engineers performing structural design for seismic effects to stay informed as to the most current design data.

TABLE 7.19
Subsurface Profile for Memphis Ground Motion Simulation (Atkinson & Beresnev, 2002)

Layer	Thickness (m)	Shear Wave Velocity (m/s)	Density (g/cm³)
1 – Alluvium	7	360	1.9
2 – Alluvium	5	360	2.0
3 – Alluvium	15	360	2.1
4 – Loess	9	360	2.2
5 – Fluvial Deposits	8	360	2.0
6 – Jackson Formation	47	520	2.1
7 – Memphis Sand	246	667	2.3
8 – Wilex Group	83	733	2.4
9 – Midway Group	580	820	2.5
10 – Paleozoic Rock	500	3,280	2.5
11	8,000	3,600	2.7
12	10,000	3,700	2.9
13	20,000	4,200	3.0

7.8 SYNTHETIC AND ARTIFICIAL GROUND MOTIONS

In addition to earthquake ground motion records obtained from online ground motion databases, synthetic and artificial records may, at times, be used in structural analysis.

Synthetic records are real records which have been modified by means other than simple amplitude scaling. Artificial records are acceleration histories generated from physics-based seismological models.

Atkinson and Beresnev (2002) produced artificial records for Memphis and St. Louis in the New Madrid seismic zone (NMSZ) from M_W 7.5 to M_W 8.0 simulations. Table 7.19 summarizes the subsurface profile used in the simulations for Memphis, Tennessee.

Fernandez and Rix (2006) produced synthetic records for the NMSZ at various location in the Mississippi Embayment. Figure 7.45 identifies seven sites within the NMSZ included in the Georgia Tech study, and further subdivides the Mississippi Embayment into “Uplands” and “Lowlands” regions. The “Lowlands” region is characterized by lower shear wave velocities in the upper 100 m, with shear wave velocities similar to those for the “Uplands” region below 100 m.

Simulation techniques have been used to develop artificial ground motion records for eastern and western Canada with magnitudes ranging from 7.0 to 9.0. A subset of these is available online (Atkinson, 2009).

Work at the University of Memphis in collaboration with the Tennessee Department of Transportation produced a set of synthetic ground motion histories appropriate for various regions across Tennessee, including west Tennessee in the New Madrid seismic zone.

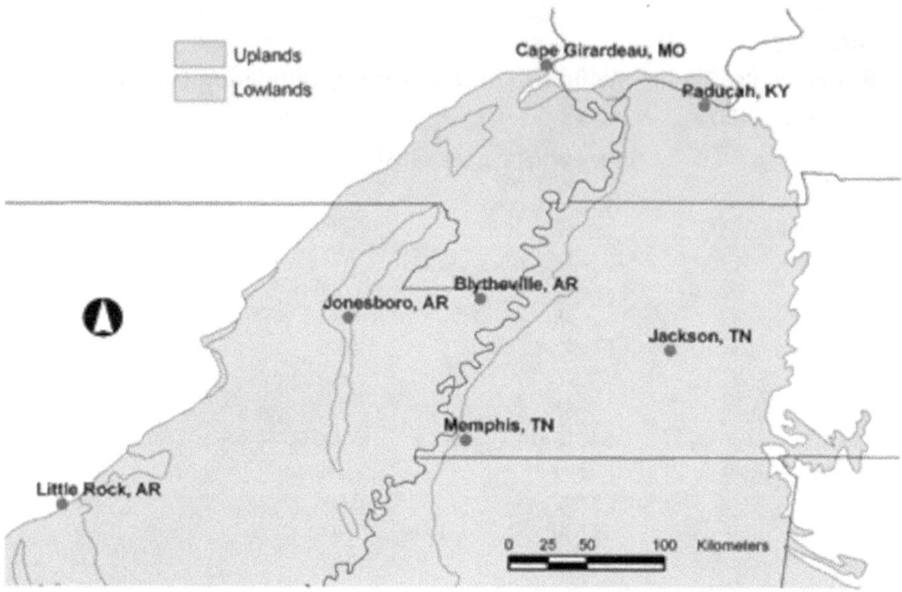

FIGURE 7.45 Georgia tech sites (Fernandez & Rix, 2006).

SeismoArtif (SeismoSoft, 2012) provides functionality for generating artificial and synthetic ground motion records.

Synthetic and artificial accelerograms must be used with caution since such procedures may produce records with unusual resulting ground motion parameters when compared to those typical of earthquake ground motion recordings from actual events. Figure 7.46 shows two artificial records generated from the same computer code with identical target spectra. Record #1 has PGA = 0.404, PGV = 241 cm/s, and PGD = 591 cm. Record #2 has PGA = 0.431 g, PGV = 54.1 cm/s, and PGD = 29.0 cm. The use of Record #1 in ground motion parameter estimation or in response history analysis of a structure would be of questionable merit.

7.8.1 Example Problem: Artificial Ground Motion Generation

For this example, suppose a one-directional analysis is required for a two-dimensional frame structure at a location defined by a design, target response spectrum with:

$$A_S = 0.366\ \text{g} \quad S_{DS} = 0.902\ \text{g} \quad S_{D1} = 0.623\ \text{g}$$

For this example, SeismoArtif is used with the *Artificial Accelerogram Generation* feature applied. To help in matching the response spectra of the artificially generated motions to the target response spectrum over a wide period range, a relatively long duration of 300 seconds is specified in the software. The period range selected for this example is extremely wide, 0.10–12.5 seconds. Seldom would this wide range be necessary.

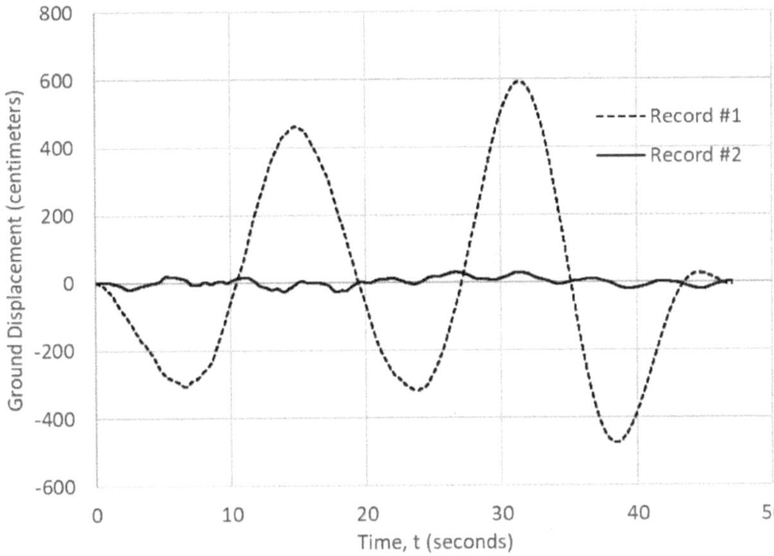

FIGURE 7.46 Example artificial accelerograms.

Figure 7.47 shows one of the eight accelerograms generated and Figure 7.48 shows the target response spectrum along with the average response spectrum of the eight accelerograms.

It would not likely be permissible in modern codes to use artificial accelerograms exclusively on a project. However, it may at times prove warranted to supplement a suite of modified ground motion records with a few artificial records to obtain a require suite size.

7.9 GROUND MOTION PROBABILITY OF EXCEEDANCE

Design ground motions are typically probabilistic in nature. A Poisson probability distribution, in which past occurrences have no effect on the likelihood of future occurrences, is often assumed. For the Poisson distribution, the probability of exceedance (PE), exposure time (t), and mean recurrence interval (MRI) are related as given by Equation 7.40:

$$\mathrm{PE} = 1 - e^{-\frac{t}{\mathrm{MRI}}} \tag{7.40}$$

Suppose that one specification for buildings requires that the design earthquake ground motion have a 2% PE in 50 years while a second specification for bridges requires design ground motion having 7% PE in 75 years. Each of these exceedance probabilities has been used in historic design specifications at some time or another. What are the mean recurrence intervals for these two design basis ground motions?

Rearrange the equation to solve for MRI:

$$\mathrm{MRI} = \frac{-t}{\ln(1 - \mathrm{PE})}$$

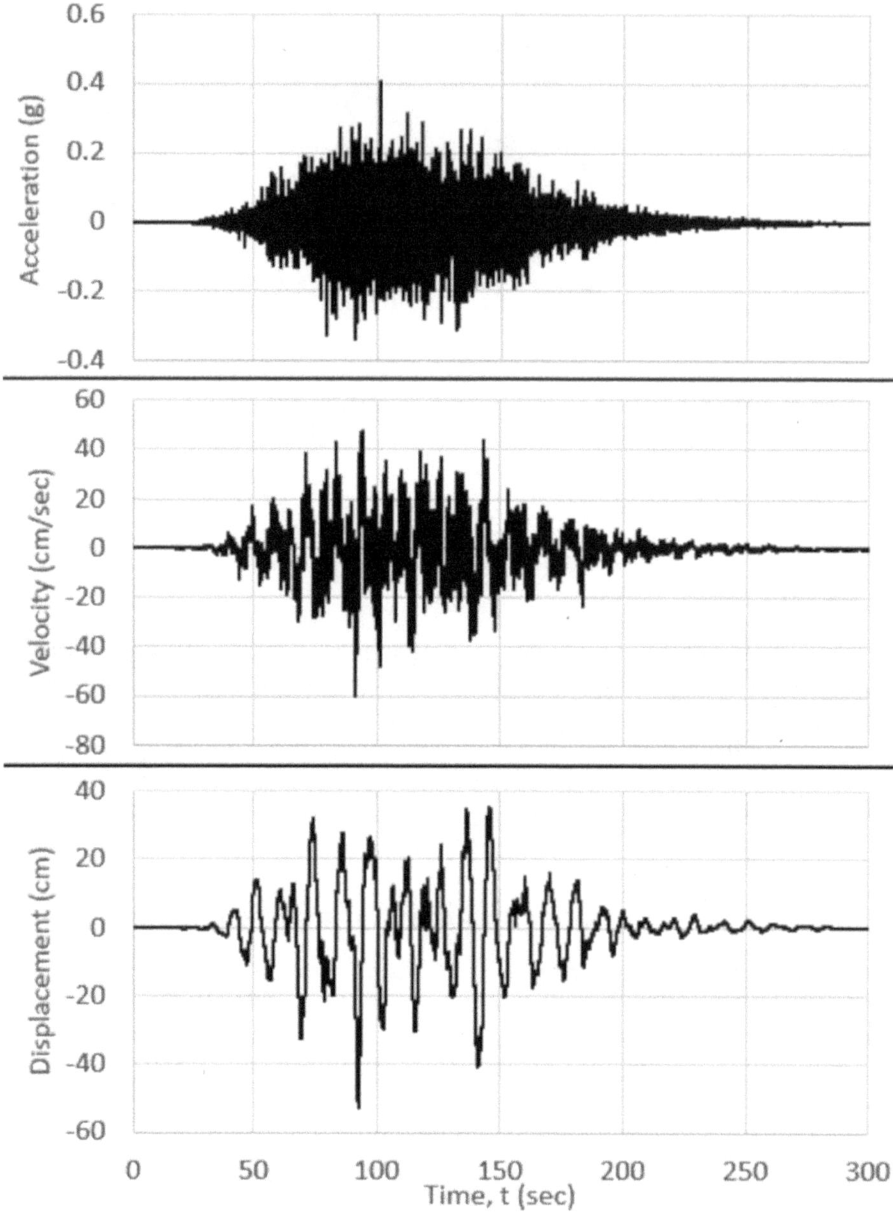

FIGURE 7.47 Artificial accelerogram generated for example problem.

So, the first case,

$$\text{MRI} = \frac{-50}{\ln(1-0.02)} = 2{,}475 \text{ years} \approx 2{,}500 \text{ years}$$

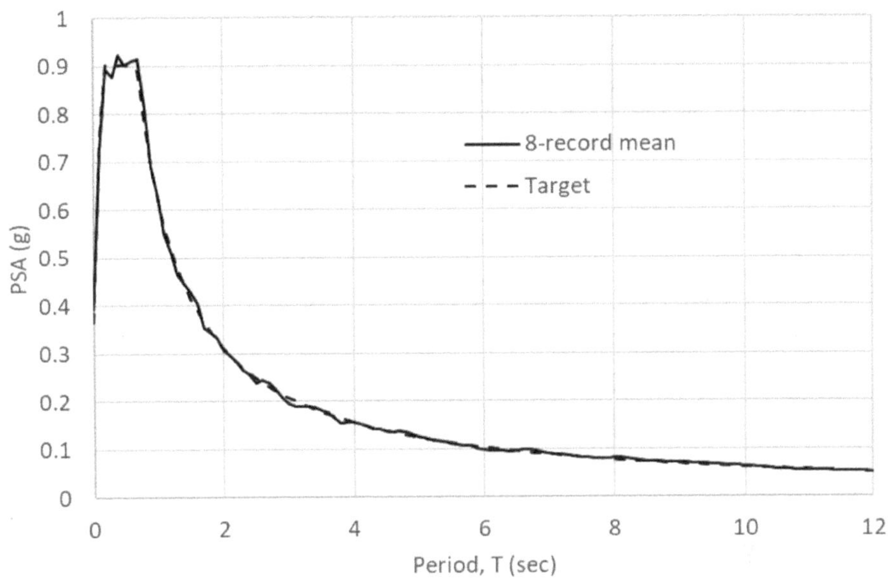

FIGURE 7.48 Target and artificial suite spectra for example problem.

And, for the second case,

$$\text{MRI} = \frac{-75}{\ln(1-0.07)} = 1,033 \text{ years} \approx 1,000 \text{ years}$$

It may appear that buildings have been historically designed for a more severe seismic loading than bridges. However, since a factor of two-thirds is applied to the 2% PE/50-Year PSA values in ASCE 7, this is not necessarily so. Two-thirds of the 2,500-Year MRI ground shaking may be greater than, about equal to, or less than the 1,000-Year MRI ground shaking. Note that the annual probability of exceedance, λ, is the inverse of MRI.

When a common exceedance probability is used across all periods, the resulting ground motion is said to be uniform hazard in nature. Modern specifications are becoming more frequently based on so-called risk-targeted ground motions, with uniform probability of collapse across geographic regions, rather than uniform hazard based. Nevertheless, the concept of exceedance probability remains an important on in design ground motion determination.

7.10 GROUND MOTION MODELS

Ground motion models (formerly referred to as attenuation equations) predict some ground motion parameter as a function of earthquake magnitude, subsurface conditions at the site in question, fault type, source-to-site distance, and possible many additional parameters, depending on the complexity of the model.

Among models available publicly are the following:

1. The Pacific Earthquake Engineering Research (PEER) center developed a ground motion model for shallow crustal earthquakes within active tectonic regions. https://peer.berkeley.edu/research/data-sciences/databases
2. A ground motion model for earthquakes within the Central and Eastern United States (CEUS) resides at the B. John Garrick Institute for the Risk Sciences: https://www.risksciences.ucla.edu/nhr3/ngaeast-gmtools
3. A ground motion model for subduction zone earthquakes resides at the B. John Garrick Institute for the Risk Sciences: https://www.risksciences.ucla.edu/nhr3/nga-subduction
4. A more recent model for shallow crustal earthquakes is available in the literature (Graizer & Kalkan, 2015) and includes estimates of mean (geometric mean) and maximum-direction (*RotD100*) spectra as well as estimates of log-based standard deviation.

7.10.1 Example Problem: Ground Motion Models

Figure 7.49 was obtained from the NGA-East GMM model for three scenarios – (1) a magnitude M_W 7.55 earthquake with a source-to-site distance equal to 55 km, (2) a magnitude M_W 6.3 earthquake with a source-to-site distance equal to 30 km,

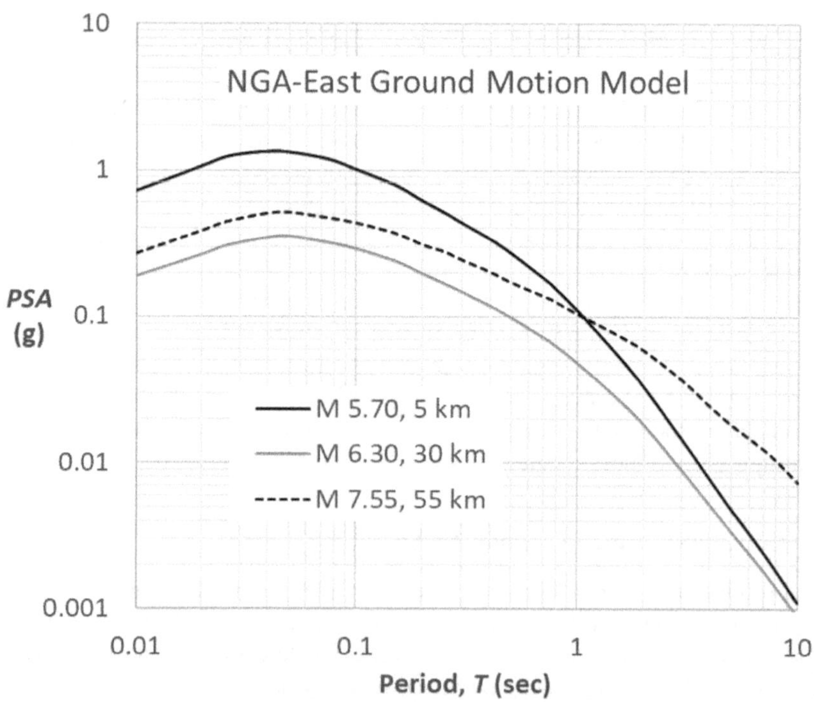

FIGURE 7.49 Hypothetical NGA-east ground motion model results.

and (3) a magnitude M_W 5.7 earthquake with a source-to-site distance equal to 5 km. The NGA-East GMM predicts ground motion parameters for subsurface conditions characterized by a shear wave velocity, V_{S30} = 3,000 m/s. Figure 7.49 illustrates an important concept: small-magnitude events close to a site may control short-period response, while large magnitude, distant events control longer period response. This is certainly not a rule for all sites but is not an uncommon phenomenon.

The USGS incorporates many GMMs into the national seismic hazard mapping project upon which most design codes and specifications are based. Logic trees are used to establish probabilistic ground motion parameters for design.

7.11 AASHTO LRFD BRIDGE DESIGN SPECIFICATIONS SEISMIC LOADING

The most current seismic loading definition in design specifications of the American Association of State Highway and Transportation Officials (AASHTO) corresponds to a 22-point design response spectrum with site class already incorporated into the United States Geological Survey (USGS) data. This eliminates the need for site factors unless a site-specific hazard analysis is warranted. The data for a given site is accessible through online, web-based tools.

https://earthquake.usgs.gov/ws/designmaps/aashto-2023/

Risk-targeted ground motions correspond to a uniform risk of collapse across all regions as opposed to a uniform hazard, as adopted in previous design specifications. The risk-targeted design basis is a 1% probability of collapse in 50 years.

While both bridge design and building design have adopted 22-point risk-targeted response spectra for design, it is important to note that ASCE 7 DRS are *RotD100* while AASHTO DRS are *GeoMean*, though both are risk-targeted. The nature of the design response spectrum in modern codes is dynamic and it is essential that engineers keep abreast of the changing bases of design seismic loading.

For ground motion selection and modification procedures necessary to perform response history analysis of bridges, the AASHTO LRFD Bridge Design Specifications (AASHTO, 2020) are relatively silent. Most bridges are designed using a modal response spectrum analysis approach in which the loading is defined entirely in terms of a design response spectrum, with no ground motion selection required.

The AASHTO Guide Specification for LRFD Seismic Bridge Design (LRFD-GS) (AASHTO, 2011) does include a section (3.4.4) on acceleration time histories for nonlinear inelastic dynamic analysis. Only three record pairs "scaled to the approximate level of the design response spectrum in the period range of significance are required". With only three record pairs, the design value of engineering parameters is to be taken as the maximum of the three records. With seven record pairs, the design action may be taken as the average of the seven analyses. Provisions in the AASHTO Guide Specification for Seismic Isolation Design (AASHTO, 2014) are virtually identical to those from the LRFD-GS.

7.11.1 Example Problem: AASHTO DRS Definition and Ground Motion Selection

Develop both the 2-point and the 22-point *GeoMean*, risk-targeted design response spectra (DRS) for a bridge site on Interstate 155 over the Mississippi River. The site coordinates are:

- latitude 36.118°
- longitude −89.616°

The inferred shear wave velocity in the upper 30 m of the subsurface profiles is 299 m/s (980 ft/s). This was determined using the OpenSHA Site Data Application (Field et al., 2003). This categorizes the bridge in a Site Class D location. Most recent codes would require that a range of shear wave velocities be considered in site classification, given that that inferred, rather than measured, values have been used. Nonetheless, for this example, assume that the true site classification is indeed D.

Using the DRS generated for the site, select a suit of 14 ground motion record pairs to be used in nonlinear response history analysis of the bridge. Use a target ln-based standard deviation of 0.60 across the period range of interest.

The fundamental periods of the bridge have been estimated from an initial design for non-seismic effects to be 1.75 seconds in the longitudinal direction and 1.20 seconds in the transverse direction. At a period of 0.22 seconds, 90% mass participation has been achieved from a modal analysis of the preliminary design. The embayment depth is estimated to be 750 m at the site.

The period range of interest will be taken from ASCE 7-22 criteria:

$$T_1 = \text{Min}(0.22, 0.2 \times 1.20) = 0.22 \text{ seconds}$$

$$T_2 = 2.0 \times 1.75 = 3.50 \text{ seconds}$$

$$T_{\text{Range}} = [0.22 - 3.50] \text{ seconds}$$

Use the USGS web-based tools to deaggregate the hazard at the site. Use the 2,500-year mean recurrence interval for the deaggregation.

https://earthquake.usgs.gov/hazards/interactive/

For peak ground acceleration (PGA), the modal (M, R) pair from deaggregation is:

$$(M, R)_{\text{PGA}} = (7.52, 13 \text{ km})$$

Other pairs contributing at least 10% to the hazard:

$$(M, R)_{\text{PGA}} = (7.70, 30 \text{ km})$$

$$(M, R)_{\text{PGA}} = (7.50, 10 \text{ km})$$

For 0.2-second PSA (S_S), the modal (M, R) pair from deaggregation is:

$$(M, R)_{SS} = (7.52, 13 \text{ km})$$

Other pairs contributing at least 10% to the hazard:

$$(M, R)_{SS} = (7.70, 30 \text{ km})$$

$$(M, R)_{SS} = (7.50, 10 \text{ km})$$

For 1.0-second PSA (S_1), the modal (M, R) pair from deaggregation is:

$$(M, R)_{S1} = (7.52, 13 \text{ km})$$

Other pairs contributing at least 10% to the hazard:

$$(M, R)_{S1} = (7.70, 30 \text{ km})$$

$$(M, R)_{S1} = (7.70, 10 \text{ km})$$

$$(M, R)_{S1} = (7.50, 10 \text{ km})$$

$$(M, R)_{S1} = (7.90, 10 \text{ km})$$

Deaggregation at a period of 2 seconds gives results similar to those for a period of 1 second.

For ground motion selection and modification, consider only records from events having magnitude in the range [7.2–8.3]. This is a relaxed range given that spectral shape will be considered in this example.

Since the site is a deep soil site, alternative site factors (not the generic, code-based site factors) were used to develop a 2-point DRS. Site factors from (Moon et al., 2016) were used. Figure 7.50 shows both the 22-point DRS and the 2-point DRS, along with the periods defining the period range of interest, T_1 and T_2. The 2-point DRS will be used as the target for ground motion selection and modification in this example.

An initial set of 110 candidate records was assembled with the desired magnitude and shear wave velocity criteria. It typically will require an initial set of candidates this large, or larger, in order to develop an acceptable suite of record pairs.

SigmaSpectra (Kottke & Rathje, 2012) was then used to generate scale factors and five 14-record-pair suites. The suite with the lowest deviation (MSE) from the target response spectrum and lowest deviation from the target ln-based standard deviation was selected. Table 7.20 shows the pertinent data for the record suite. The ensuing Figure 7.51 depicts the match to the target DRS and the target σ.

Records from both subduction and shallow crustal earthquakes were considered for this hypothetical problem. This may not always be acceptable, and engineers should clearly define the governing criteria up front when such procedure are required to select and modify ground motion records for NLRHA (or LRHA for that matter).

No more than 4 record pairs (30% of the suite size) were selected from any individual earthquake. This is generally considered to be good practice, though not

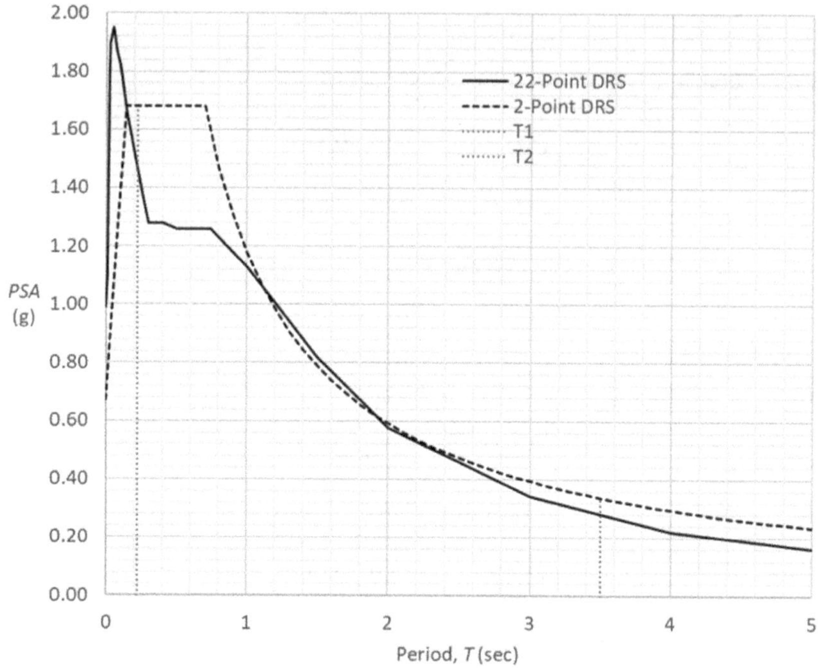

FIGURE 7.50 Design response spectra for AASHTO DRS example.

TABLE 7.20
Metadata for Ground Motion Suite – AASHTO Example

EQ	Year	M_W	RSN	Station	Scale	V_{S30}, m/s
Central Chile	1985	7.98	6001750	LLAYLLAY	3.300	270
Central Chile	1985	7.98	6001750	LLAYLLAY	3.300	270
El Mayor	2010	7.20	5823	Chihuahua	3.140	242
El Mayor	2010	7.20	5823	Chihuahua	3.140	242
El Mayor	2010	7.20	5988	Meloland E Holton	2.808	196
El Mayor	2010	7.20	5988	Meloland E Holton	2.808	196
El Mayor	2010	7.20	8606	Westside Elem.	3.702	242
El Mayor	2010	7.20	8606	Westside Elem.	3.702	242
SMART1(45)	1986	7.30	582	O-08	3.458	357
SMART1(45)	1986	7.30	582	O-08	3.458	357
Landers	1992	7.29	832	Amboy	2.658	383
Landers	1992	7.28	832	Amboy	2.658	383
Chi-Chi	1999	7.62	1203	CHY036	4.120	233
Chi-Chi	1999	7.62	1203	CHY036	4.120	233
Chi-Chi	1999	7.62	1236	CHY088	2.949	318
Chi-Chi	1999	7.62	1236	CHY088	2.949	318
Chi-Chi	1999	7.62	1331	ILA041	1.970	197

(*Continued*)

TABLE 7.20 (*Continued*)
Metadata for Ground Motion Suite – AASHTO Example

EQ	Year	M_W	RSN	Station	Scale	V_{S30}, m/s
Chi-Chi	1999	7.62	1331	ILA041	1.970	197
Chi-Chi	1999	7.62	1495	TCU055	3.410	359
Chi-Chi	1999	7.62	1495	TCU055	3.410	359
Pezeshk-Tdot	2015	7.70	Synthetic	Site-03-02	1.510	274
Pezeshk-Tdot	2015	7.70	Synthetic	Site-03-02	1.510	274
Pezeshk-Tdot	2015	7.70	Synthetic	Site-03–04	1.789	274
Pezeshk-Tdot	2015	7.70	Synthetic	Site-03–04	1.789	274
Pezeshk-Tdot	2015	7.70	Synthetic	Site-03–06	2.541	274
Pezeshk-Tdot	2015	7.70	Synthetic	Site-03–06	2.541	274
SMART1(45)	1986	7.30	3671	I-12	2.957	276
SMART1(45)	1986	7.30	3671	I-12	2.957	276
Average		**7.50**			**2.88**	**278.21**

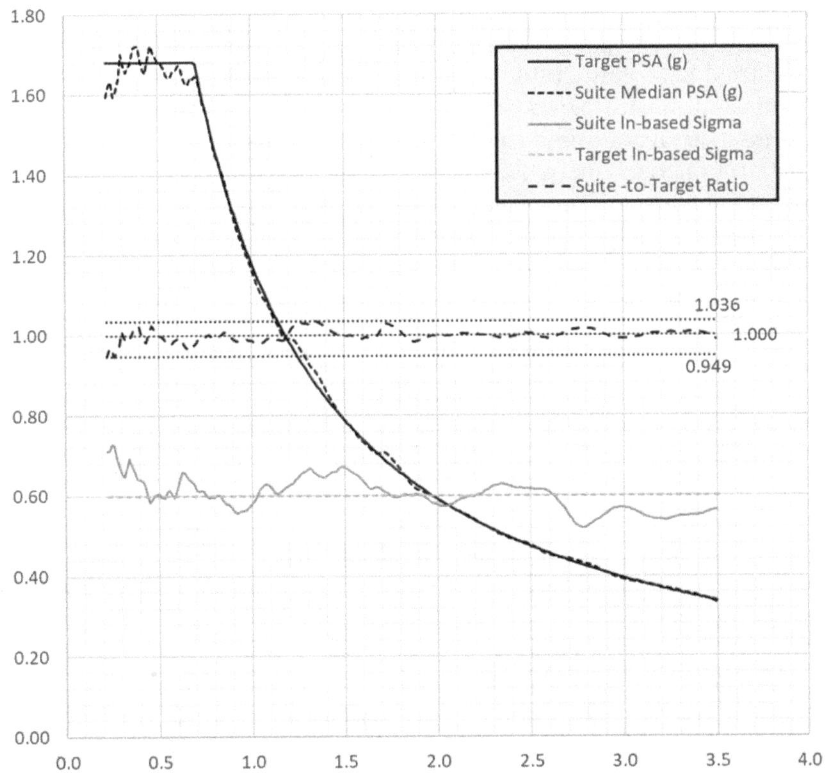

FIGURE 7.51 Suite and target spectra and variability.

always a specific requirement. No record requiring a scale factor greater than 5 was selected for the final suite as well. Real, recorded record pairs were supplemented with synthetic records generated at the University of Memphis for the Tennessee Department of Transportation in 2015.

The Figure shows the following characteristics of the selected 14-record-pair suite.

- The maximum suite mean-to-target ratio within the period range of interest is 1.036.
- The minimum suite mean-to-target ratio within the period range of interest is 0.949.
- The average suite mean-to-target ratio within the period range of interest is 1.000.
- The average ln-based standard deviation of the 14-record-pair suite within the period range of interest, in this case, is exactly equal to the target value of 0.600, with a range of values from 0.518 to 0.729.

See also Section 7.14.8 for additional discussions on seismic design criteria for bridges.

7.12 ASCE 43-19 SEISMIC LOADING

Seismic loading for components in nuclear facilities is defined in ASCE 43 (American Society of Civil Engineers, 2019). At any given period, the design response spectrum (DRS) ordinate is equal to the product of a period-dependent scale factor, SF, and the uniform hazard spectral ordinate corresponding to an exceedance frequency equal to H_P. Table 7.21 summarizes the required exceedance frequencies and Equations 7.41 through 7.46 summarize the relationships needed to develop the DRS for a given site. The parameters are dependent on the assigned seismic design category (SDC), which should not be confused with the term "seismic design category" used in other specifications. If the DRS at $T=0.0$ seconds is less than PGA_{Min}, then the entire DRS is to be scaled up.

$$\text{DRS} = \text{SF} \cdot \text{UHRS}_{\text{HP}} \tag{7.41}$$

$$A_R = \frac{\text{SA}_{HP}}{\text{SA}_{HD}} \tag{7.42}$$

$$\text{SF} = \text{Max}\left\{\text{SF}_1,\ \text{SF}_2,\ \text{SF}_3\right\} \tag{7.43}$$

TABLE 7.21
DRS Parameters in ASCE 43-19

SDC	$H_P=P_F$	$H_D=10\times P_F$	MRI_P	MRI_D	PGA_{min}
2	4×10^{-4}	4×10^{-3}	2,500 years	250 years	0.04 g
3	1×10^{-4}	1×10^{-3}	10,000 years	1,000 years	0.06 g
4	4×10^{-5}	4×10^{-4}	25,000 years	2,500 years	0.08 g
5	1×10^{-5}	1×10^{-4}	100,000 years	10,000 years	0.10 g

$$\mathrm{SF}_1 = 1.0(A_R)^{-1.0} \tag{7.44}$$

$$\mathrm{SF}_2 = 0.6(A_R)^{-0.20} \tag{7.45}$$

$$\mathrm{SF}_3 = 0.45 \tag{7.46}$$

ASCE 43 requires that ground motion record suites for design have a time step no larger than 0.01 seconds and be matched to the target response spectrum over a frequency range no less restrictive than 0.2–25 Hz. The suite average spectra may fall below the target spectrum no more than 10% at no more than 9 consecutive frequency points. Frequency points are to be assigned at a minimum of 100 points per frequency decade on a logarithmic scale basis. Further, the suite average spectrum may not exceed the target by more than 30% at any frequency within the specified range.

7.12.1 Example Problem: ASCE 43 DRS Calculations

For this example, it has been assumed that a 22-point design response spectrum based on the 2018 update of the USGS seismic data is an acceptable option for a project in Hanford, Washington (a former nuclear operation site). There will likely be occasions when a site-specific study is required for nuclear facilities.

Develop the DRS for a Seismic Design Category 5 (SDC5) project in Hanford, Washington in accordance with ASCE 43 provisions. Compare this DRS to that for a Risk Category IV structure from ASCE 7-22.

For SDC5, $H_D = 1 \times 10^{-4}$ and $H_P = 1 \times 10^{-5}$, from Table 1.21. The coordinates of the project site are 46.534 degrees North, 119.422° West (46.534, −119.422). Western longitudes are negative by modern convention.

For this example, the Dynamic Curves option at the USGS Earthquake Hazards Toolbox (United States Geological Survey, 2023) will be used to obtain uniform hazard ground motion data for the project site. In addition, an inferred shear wave velocity will be assumed accurate for the site. Using OpenSHA (Field et al., 2003), an inferred shear wave velocity of 315 m/s = 1,033 fps has been obtained.

To account for uncertainty in inferred (versus measured) shear wave velocity, a factor of 1.3 will be applied to the inferred value to obtain a range of shear wave velocities equal to:

$$[1{,}033/1.3 = 795 \text{ fps}, \quad 1{,}033 \times 1.3 = 1{,}343 \text{ fps}]$$

Site Classes CD ($1{,}000 < V_{S30} < 1{,}450$ fps) and D ($700 < V_{S30} < 1{,}000$ fps) are included in the range of shear wave velocities obtained. When soil properties are not known in sufficient detail, ASCE 7-22, Section 20.1 requires that the more severe of Site Classes C, CD, and D be used at each period. For this example, the assumption is made that an inferred shear wave velocity does not constitute adequate detail and the more severe of Site Classes C, CD, and D at each period will be adopted.

The USGS Toolbox is located here: https://earthquake.usgs.gov/nshmp/

OpenSHA applications may be found at: https://opensha.org/

From the USGS toolbox website, the data in Table 7.22 have been obtained. Parenthetical values represent the controlling Site Class.

The data tabulated above contain the 10,000-year mean recurrence interval (H_D-based) and the 100,000-year mean recurrence interval pseudo-spectral acceleration values for the 22-point design response spectrum from ASCE 7. Both geometric mean (of two horizontal components) and maximum-direction (*RotD100*) values are provided, as well as the ratio *RotD100/GeoMean*. This example illustrates, not only the steps in developing a DRS for ASCE 43, but also the conversion from *GeoMean* to *RotD100* spectra. See Section 7.5 for ASCE 7-22 conversion from *GeoMean* to *RotD100* spectra.

The data tabulated above do not represent the ASCE 43-19 DRS, merely the first step in developing the ASCE 43-19 DRS for this example. Table 7.22 represents uniform hazard spectra and ASCE 43-19 spectra are not uniform hazard based. At each period, the scale factor must now be computed, and the DRS spectrum generated using Equations 7.41 through 7.46. The resulting *RotD100*-based DRS is tabulated in Table 7.23 and depicted graphically in Figure 7.52. For design requiring the *GeoMean*-based, rather than the *RotD100*-based, DRS, the indicated

TABLE 7.22
Uniform Hazard Spectra for ASCE 43 Example

	Geometric Mean		*Scaled to RotD100*		*RotD100/GeoMean*	
Period, *T* (seconds)	**H_D-based PSA (g)**	**H_P-based PSA (g)**	**H_D-based PSA (g)**	**H_P-based PSA (g)**	**H_D-based PSA (g)**	**H_P-based PSA (g)**
0.000	0.431 (D)	0.873 (CD)	0.431	0.873	1.000	1.000
0.010	0.433 (D)	0.878 (CD)	0.520	1.054	1.200	1.200
0.020	0.437 (D)	0.880 (CD)	0.525	1.056	1.200	1.200
0.030	0.450 (CD)	0.926 (CD)	0.540	1.111	1.200	1.200
0.050	0.541 (C)	1.173 (C)	0.650	1.408	1.200	1.200
0.075	0.700 (C)	1.521 (C)	0.840	1.825	1.200	1.200
0.100	0.826 (C)	1.790 (C)	0.991	2.148	1.200	1.200
0.150	0.973 (CD)	2.036 (C)	1.167	2.443	1.200	1.200
0.200	1.039 (CD)	2.108 (CD)	1.247	2.530	1.200	1.200
0.250	1.081 (D)	2.154 (CD)	1.301	2.591	1.203	1.203
0.300	1.093 (D)	2.140 (CD)	1.319	2.581	1.206	1.206
0.400	1.024 (D)	2.065 (D)	1.243	2.505	1.213	1.213
0.500	0.942 (D)	1.952 (D)	1.148	2.380	1.219	1.219
0.750	0.710 (D)	1.538 (D)	0.876	1.898	1.234	1.234
1.000	0.564 (D)	1.255 (D)	0.706	1.568	1.250	1.250
1.500	0.383 (D)	0.854 (D)	0.479	1.070	1.253	1.253
2.000	0.291 (D)	0.634 (D)	0.365	0.796	1.256	1.256
3.000	0.201 (D)	0.416 (D)	0.253	0.524	1.261	1.261
4.000	0.158 (D)	0.322 (D)	0.200	0.408	1.267	1.267
5.000	0.134 (D)	0.285 (D)	0.171	0.362	1.272	1.272
7.500	0.100 (D)	0.230 (D)	0.129	0.295	1.286	1.286
10.000	0.086 (D)	0.215 (D)	0.112	0.279	1.300	1.300

TABLE 7.23
***RotD100*-based Design Response Spectrum for ASCE 43 Example**

Period, T (seconds)	A_R	SF1	SF2	SF	DRS PSA (g)
0.000	2.025	0.494	0.521	0.521	0.455
0.010	2.026	0.494	0.521	0.521	0.549
0.020	2.013	0.497	0.522	0.522	0.551
0.030	2.057	0.486	0.519	0.519	0.577
0.050	2.167	0.461	0.514	0.514	0.724
0.075	2.173	0.460	0.514	0.514	0.937
0.100	2.167	0.461	0.514	0.514	1.104
0.150	2.093	0.478	0.518	0.518	1.264
0.200	2.028	0.493	0.521	0.521	1.318
0.250	1.992	0.502	0.523	0.523	1.354
0.300	1.957	0.511	0.525	0.525	1.354
0.400	2.016	0.496	0.522	0.522	1.306
0.500	2.073	0.482	0.519	0.519	1.234
0.750	2.167	0.462	0.514	0.514	0.975
1.000	2.223	0.450	0.511	0.511	0.802
1.500	2.232	0.448	0.511	0.511	0.547
2.000	2.178	0.459	0.513	0.513	0.409
3.000	2.070	0.483	0.519	0.519	0.272
4.000	2.043	0.489	0.520	0.520	0.212
5.000	2.118	0.472	0.516	0.516	0.187
7.500	2.298	0.435	0.508	0.508	0.150
10.000	2.496	0.401	0.500	0.500	0.139

backward conversions could be made. According to ASCE 7-22, conversion between *GeoMean*-based and *RotD100*-based design response spectra shall adhere to the following rules:

- For periods less than or equal to 0.20 seconds, $PSA_{RotD100} = 1.2\ PSA_{GeoMean}$.
- For a period of 1.0 seconds, $PSA_{RotD100} = 1.25\ PSA_{GeoMean}$.
- For periods greater than or equal to 10 seconds, $PSA_{RotD100} = 1.30\ PSA_{GeoMean}$.
- For periods in the range (0.20–1.00) seconds, use a linear interpolation.
- For periods in the range (1.00–10.0) seconds, use a linear interpolation.

7.13 ASCE 41-17 SEISMIC LOADING

For the evaluation and retrofit of existing buildings, seismic loading in ASCE 41-17 (ASCE, 2017) includes design response spectral ordinates for various hazard levels. The hazard levels are defined as follows:

- **BSE-1E:** Basic Safety Earthquake 1 for use with basic performance objectives for existing buildings; the design response spectrum is that corresponding to a 20% probability of exceedance in 50 years with a

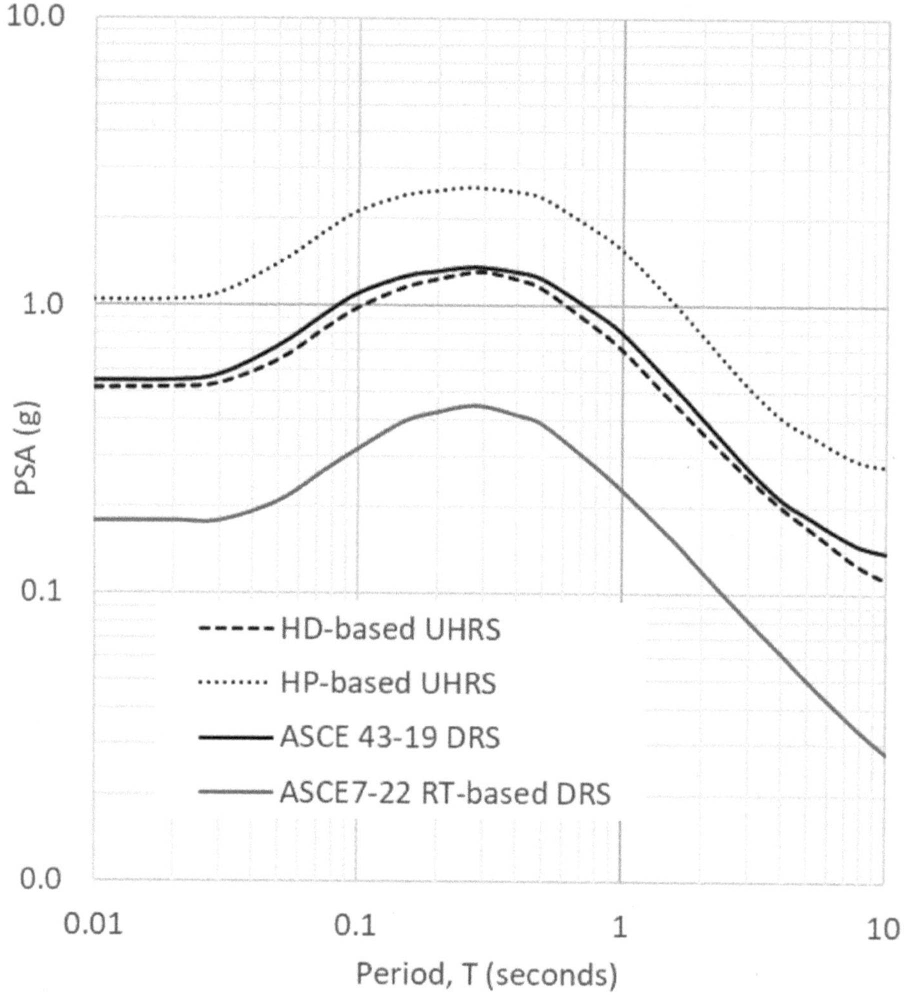

FIGURE 7.52 *RotD100*-based DRS for ASCE 43 example.

maximum-direction (*RotD100*) basis. However, the BSE-1E design ground motion need not exceed the BSE-1N ground motion.

- **BSE-2E:** Basic Safety Earthquake 2 for use with basic performance objectives for existing buildings; the design response spectrum is that corresponding to a 5% probability of exceedance in 50 years with a maximum-direction (*RotD100*) basis. However, the BSE-2E design ground motion need not exceed the BSE-2N ground motion.
- **BSE-1N:** Basic Safety Earthquake 1 for use with basic performance objectives equivalent to new building standards; the design response spectrum is two-thirds of that for BSE-2N.

- **BSE-2N:** Basic Safety Earthquake 1 for use with basic performance objectives equivalent to new building standards; the design response spectrum is that corresponding to the ground motion defined in Section 11.4 of ASCE 7, the risk-targeted, maximum-direction (*RotD100*) DRS corresponding to a 1% probability of collapse in 50 years. This design ground motion for the BSE-2N hazard level is consistent with the maximum considered, risk-targeted ground motion (MCE_R) from ASCE 7.

7.14 STRUCTURAL CRITERIA FOR EARTHQUAKE LOADING

While the focus of this book is on the determination of extreme loads on structures, it seems appropriate to make mention of analytical techniques used to assess the effects of such loadings, without going into detail. Also beneficial will be discussions on design concepts such as capacity-protection principles. For excellent and exhaustive treatments on structural analysis, references such as Chopra (2016) and Charney et al. (2020) should prove valuable options.

One important consideration in the specification of extreme loading is the concept of a risk category in ASCE 7-22. The proper selection of a risk category requires consideration of the consequences of failure.

Risk Category I: buildings and other structures that represent low risk to human life in the event of failure; examples include barns and storage facilities; the seismic importance factor $I_e = 1.00$ for Risk Category I structures.

Risk Category II: all buildings and other structures not in Category I, III, or IV; most residential, commercial, and industrial buildings are Risk Category II structures; the seismic importance factor, $I_e = 1.00$ for Risk Category II structures.

Risk Category III: buildings and other structures, the failure of which could pose a substantial risk to human life; examples include theaters, lecture halls, elementary schools, prisons, and small health-care facilities; the seismic importance factor $I_e = 1.25$ for Risk Category III buildings; risk category III includes buildings and other structures, not in Risk Category IV, with potential to cause a substantial economic impact and/or mass disruption of day-to-day civilian life in the event of failure including, but not limited to:

> facilities dealing with hazardous fuels, hazardous chemicals, hazardous waste, or explosives where the quantity of material exceeds a threshold established by the Authority Having Jurisdiction and where the quantity is sufficient to pose a threat to the public if released.

Risk Category IV: buildings and other structures designated as essential facilities are intended to remain operational in the event of extreme environmental loading from flood, wind, snow, or earthquakes; examples include hospitals, police stations, emergency communications centers; the seismic importance factor $I_e = 1.50$ for Risk Category IV structures.

AASHTO, on the other hand, categorizes bridges as critical, recovery, or ordinary, with varying criteria for each category.

The permissible analysis methods in most codes vary with the risk category. Some of the analysis methods used in seismic engineering will be discussed now.

7.14.1 Equivalent Lateral Force Procedure (ELF) for Buildings in ASCE 7

For structures designed using an equivalent lateral force procedure found in ASCE 7-16, the base shear, V, is computed from a seismic coefficient, C_s, as shown in Equations 7.47 through 7.51. The seismic coefficient depends on the natural period, T, of the structure and the spectral acceleration, S_a, at that period. Equation 7.49 gives the upper limit on C_s, while Equation 50 gives a lower limit on the value used for design. The lower limit defined by Equation 7.51 applies only in cases where S_1 is greater than or equal to 0.60 g.

Given that the natural periods of the structure are unknown prior to design completion. Equation 7.52 provides an estimate, T_a, of the fundamental natural period, T. Final design should incorporate the actual fundamental period as opposed to an approximate from the Equation and may require re-computation of the seismic loads. In Equation 7.52, h_n is the structural height and values for C_t and x from ASCE 7 include:

- steel moment frames: $C_t = 0.028$ and $x = 0.80$
- concrete moment frames: $C_t = 0.016$ and $x = 0.90$
- steel eccentrically braced frames: $C_t = 0.030$ and $x = 0.75$
- steel buckling-restrained braced frames: $C_t = 0.030$ and $x = 0.75$
- all other structural systems: $C_t = 0.020$ and $x = 0.75$

While ASCE 7 does require that the natural period be computed from appropriately modeled mass and stiffness, there is a limit on the maximum period that may be used for the ELF method. The maximum permitted period to be used in the ELF method is the product of C_u and the approximated period, T_a, from Equation 7.52. The coefficient, C_u, depends on the 1-second design spectral acceleration, S_{D1}:

- for $S_{D1} \geq 0.40$, $C_u = 1.4$
- for $S_{D1} = 0.30$, $C_u = 1.4$
- for $S_{D1} = 0.20$, $C_u = 1.5$
- for $S_{D1} = 0.15$, $C_u = 1.6$
- For $S_{D1} \leq 0.10$, $C_u = 1.7$

$$V = C_s W \tag{7.47}$$

$$C_s = \frac{S_{DS}}{\left(\frac{R}{I_e}\right)} \tag{7.48}$$

$$C_s \leq \begin{cases} \dfrac{S_{D1}}{T\left(\dfrac{R}{I_e}\right)} & \text{for } T \leq T_L \\ \dfrac{S_{D1}T_L}{T^2\left(\dfrac{R}{I_e}\right)} & \text{for } T > T_L \end{cases} \tag{7.49}$$

$$C_s \geq 0.044 S_{DS} I_e \geq 0.01 \tag{7.50}$$

$$C_s \geq \frac{0.5 S_1}{\left(\frac{R}{I_e}\right)} \tag{7.51}$$

$$T_a = C_t h_n^x \tag{7.52}$$

Once the base shear has been determined, it must be distributed horizontally to the various lateral force-resisting bays, and vertically to the various floor levels in each bay.

7.14.1.1 Horizontal Distribution of Seismic Shear

Horizontal distribution of the equivalent lateral base shear, V, depends on the stiffness of the lateral force-resisting system (LFRS) elements.

Figure 7.53 is a plan view of a rectangular-shaped building. Even if the theoretical center of mass coincides with the theoretical center of stiffness, some accidental torsion is often warranted due to uncertainties in the theoretical locations. An accidental torsional offset equal to 5% of the building dimension perpendicular to the direction of loading is sometimes called for in design specifications. Equations 7.53 and 7.54 give the shear in the ith LFRS element due to loads in the x and z directions, respectively. The first term on the right-hand side of the equations is the direct shear term, and only LFRS elements parallel to the direction of load receive direct shear. The second term on the right-hand side represents torsion-induced shear, and all

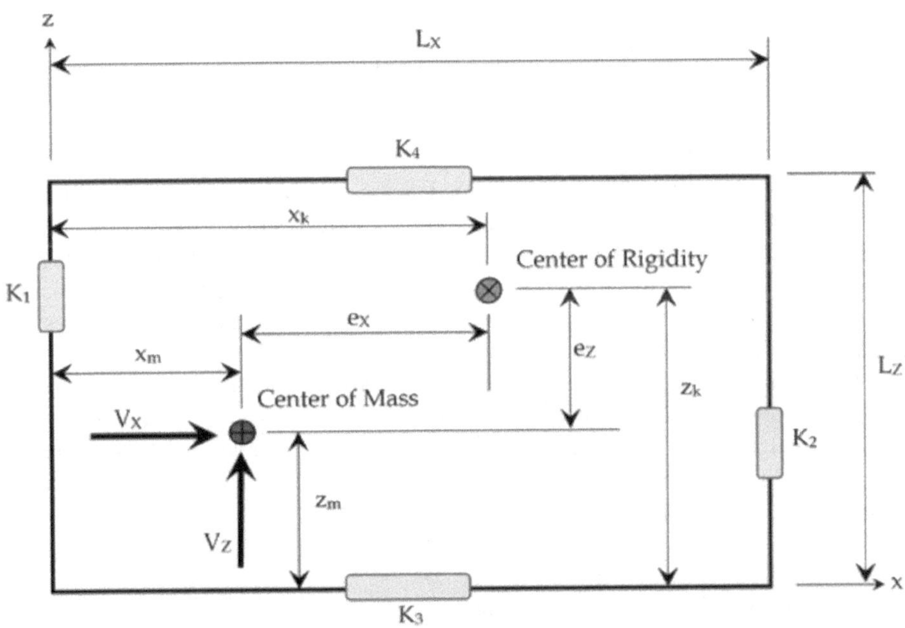

FIGURE 7.53 Horizontal distribution of LFRS elements.

LFRS elements receive torsion-induced shear. Hence, for example, the first term in Equation 7.53 depends on the stiffness in the x-direction (K_{ix}) while the second term depends on the in-plane stiffness, K_i, of any LFRS element in the system.

Whether the second term is added or subtracted depends on the location of the LFRS element with respect to the center of mass. If the center of rigidity lies between the center of mass and the element being considered, then the second term (torsional-induced shear) is subtracted from the first term (direct shear). If the center of mass and the element in question lie on the same side as the center of rigidity, then the second term is added to the first term.

It is a typical requirement that bi-directional effects be incorporated into the design. When V_x is at a maximum, V_z is assumed to be at some fraction of the maximum. This fraction is 30% in ASCE 7 (ASCE, 2022) and 40% in ASCE 43 (ASCE, 2019). In ASCE 7, the more severe of the loading combinations given in Equation 7.55 is the design shear for the stack of LFRS elements in a given bay. In ASCE 41, simply replace the 0.30 factor with 0.40 in Equation 7.55:

$$(V_i)_x = V_x\left(\frac{K_{ix}}{\sum K_{ix}}\right) \pm \frac{V_x e_z (K_i d_i)}{\sum (K_i d_i^2)} \tag{7.53}$$

$$(V_i)_z = V_z\left(\frac{K_{iz}}{\sum K_{izx}}\right) \pm \frac{V_z e_x (K_i d_i)}{\sum (K_i d_i^2)} \tag{7.54}$$

$$V_i = \text{Max}\begin{cases} 1.00(V_i)_x + 0.30(V_i)_z \\ 0.30(V_i)_x + 1.00(V_i)_z \end{cases} \tag{7.55}$$

7.14.1.2 Vertical Distribution of Seismic Shear

Once the shear in each LFRS bay has been determined, the shear in that bay must be distributed vertically to the various floor levels in a multi-story building.

Equation 7.56 provides the means of distributing the shear vertically in ASCE 7-16. The shear at a given floor level of an LFRS element, F_{xi}, is equal to the coefficient C_{vx} multiplied by the total shear for the LFRS element. Equation 7.57 defines the coefficient, C_{vx}.

The weight at each floor level, w_x, the height above ground of each floor level, h_x, and the exponent, k, all appear in Equation 7.57. Note that h_x in this equation is not the story height but the height above ground for each floor level. The exponent k depends on the fundamental period of the structure in the direction under consideration and is determined as follows:

- $k = 1.00$ for $T = 0.50$ seconds or less
- $k = 2.00$ for $T = 2.50$ seconds or more
- interpolate for T between 0.50 and 2.50 seconds:

$$F_{xi} = C_{vx} V_i \tag{7.56}$$

$$C_{vx} = \frac{w_x h_x^k}{\sum_{i=1}^{N} w_i h_i^k} \tag{7.57}$$

7.14.2 Modal Analysis and Superposition

A 2DOF system may be converted to two SDOF systems for solution to earthquake loading defined in terms of a design response spectrum (DRS). In fact, an n-dof system can be converted to n SDOF system. It is often more computationally efficient to solve n SDOF systems than a single n-dof system. The procedure is called modal superposition. The procedure requires that the eigenvalues (the squares of the natural frequencies) and eigenvectors (the mode shapes) be determined. Modal superposition is made possible by the fact that the eigenvectors are orthogonal with respect to the mass and stiffness in such systems. This orthogonality is expressed in Equations 7.58 and 7.59. GM_i is the generalized mass for mode i. GK_i is the generalized stiffness for mode i.

Equation 7.60 shows that the ratio of GK_i to GM_i is simply the eigenvalue (the square of the circular frequency) for mode i. Given that any multiple of a mode shape is still a valid mode shape, it is convenient to multiply each mode shape by a scalar quantity such that $GM_i = 1$ for each mode. This is called mass normalization of the mode shapes:

$$\phi_j^T M \phi_i = \begin{cases} GM_i, & j = i \\ 0, & j \neq i \end{cases} \tag{7.58}$$

$$\phi_j^T K \phi_i = \begin{cases} GK_i, & j = i \\ 0, & j \neq i \end{cases} \tag{7.59}$$

$$\lambda_j = \omega_{nj}^2 = \frac{GK_j}{GM_j} \tag{7.60}$$

Once the generalized modal properties have been determined, the n-dof system may be solved as n SDOF systems using modal superposition. The modal superposition relationships are given in Equations 7.61 through 7.64. For the shear building model (see Figure 7.61), the vector r is simply a vector of n 1's. The effective modal mass for mode i is EMM_i. The modal participation factor for mode i is Γ_i. SD_i is the spectral displacement for mode i and PSA_i is the pseudo-spectral acceleration for mode i:

$$EMM_i = \frac{\left[\left(\phi^T M r\right)_i\right]^2}{\left(\phi^T M \phi\right)_i} \tag{7.61}$$

$$\Gamma_i = \frac{\left(\phi^T M r\right)_i}{\left(\phi^T M \phi\right)_i} \tag{7.62}$$

$$\{u\}_i = \Gamma_i(\text{SD}_i)\phi_i \tag{7.63}$$

$$\{\ddot{u}\}_i = \Gamma_i(\text{PSA}_i)\phi_i \tag{7.64}$$

Modal superposition requires that each mode be solved individually, and then the contributions from each mode are combined. There are multiple methods used in the modal combination process. The SRSS combination rule uses the square-root-of-the-sum-of-the-squares to estimate the total response. The absolute sum method uses the sum-of-absolute-values to estimate the total response. Modern software also typically permits the complete quadratic combination (CQC) rule to be applied to modal superposition. Refer to textbooks on structural dynamics for details of the CQC method.

It is important to estimate drift at each story of a shear building for each mode individually before combining. This will be distinct from the difference in combined floor displacements at adjacent levels.

An important parameter produced in modal analysis/superposition is the percent mass participating in each mode, calculated by Equation 7.65. The sum of the effective modal masses for all modes must equal the total system mass, or computational inaccuracies exist. Many design specifications require that enough modes be considered such that at least 90% of the total system mass has been included. So, it may be that a 20-dof system only requires consideration of three modes, just to give a hypothetical example:

$$\text{PM}_i = \frac{\text{EMM}_i}{\sum_{j=1}^{n} \text{EMM}_j} \times 100 \tag{7.65}$$

7.14.3 Linear Response History Analysis

While seismic analysis of structures typically is performed through response spectrum analysis (RSA), at times, it becomes necessary to use response history analysis for earthquake-resistant design. Response history analysis may be linear (LRHA), but more often, when response history analysis is employed, the analysis directly incorporates nonlinear effects – nonlinear response history analysis (NLRHA). The term "time history analysis" is sometimes used but is actually a misnomer. The procedure produces results in the form of response history.

LRHA and NLRHA require not only a design response spectrum (DRS) but also a suite of ground motion record pairs to be applied to the base of the structure. Typically, at least seven, but more than likely 11 or more, record pairs are required. The record pairs may be from actual event or synthetically generated. In either case, some modification of the record pairs is usually necessary.

Ground motion modification procedures include (1) amplitude scaling, (2) spectral matching in the time domain, and (3) spectral matching in the frequency domain.

The preferred method is amplitude scaling in which the acceleration values for a given record are amplified by a scalar to make the response spectrum of the record closely match the DRS. This is preferred because variability is retained and is a reality.

Spectral matching in the time domain involved adding wavelets directly to the record to make the record response spectrum more closely match the DRS.

Spectral matching in the frequency domain involves generating a Fourier spectrum consistent with the DRS. The Fourier spectrum for each ground motion record is computed and adjusted to closely match the DRS Fourier spectrum. The modified record Fourier spectrum is then converted back to a new ground motion record.

Refer to Section 7.14.5 for more details on ground motion selection and modification.

Spectral matching artificially reduces variability in ground motion records. This does not mean that spectral matching should never be used. It simply means that the engineer must be well-versed in ground motion theory to properly perform spectral matching.

Damping is explicitly considered in modal superposition by specifying the design response spectrum (DRS) corresponding to a specific level of damping (usually 5% of critical). For response history analysis, the following sections discuss some of the available options in modern software.

For response history analysis, one frequently adopted model is the Rayleigh damping model, in which the damping matrix is proportional to the mass and stiffness matrices, as given by Equation 7.66. With Rayleigh damping, the mode shapes (eigenvectors) are orthogonal with respect to the resulting damping matrix. It can be shown that such a model produces a fraction of critical damping in each mode, as given by Equation 7.67. The first term in Equation 7.67 represents the mass—proportional component of damping, and the second term in Equation 7.67 represents the stiffness-proportional component of damping. The stiffness proportionality parameter, β, in Equation 7.48 should not be confused with the frequency ratio parameter, β, from the analysis of SDOF systems subjected to harmonic loading:

$$[C]=\alpha[M]+\beta[K] \tag{7.66}$$

$$\xi_i=\alpha\left(\frac{1}{2\omega_i}\right)+\beta\left(\frac{\omega_i}{2}\right)=\alpha\left(\frac{T_i}{4\pi}\right)+\left(\frac{\pi}{T_i}\right) \tag{7.67}$$

With Rayleigh damping, it is possible to control the fraction of critical damping in any two modes. Suppose we choose to specify a particular fraction of damping, ξ_{design}, for two frequencies, ω_1 and ω_2, corresponding to periods, T_1 and T_2. Then Equation 7.67 can be solved for the parameters, α and β. The resulting expressions are shown in Equations 7.68 and 7.69. Software which incorporates Rayleigh damping for NLRHA requires the user to specify the α and β parameters to be used in the analysis. Default values for such parameters should always be checked and overridden if necessary. Once α and β have been determined, the effective damping level at any other frequency can be determined by Equation 7.67:

$$\beta=\frac{2\xi_{\text{design}}}{\omega_1+\omega_2}=\frac{T_1T_2\xi_{\text{design}}}{\pi(T_1+T_2)} \tag{7.68}$$

$$\alpha=\frac{2\omega_1\omega_2\xi_{\text{design}}}{\omega_1+\omega_2}=\frac{4\pi\xi_{\text{design}}}{T_1+T_2} \tag{7.69}$$

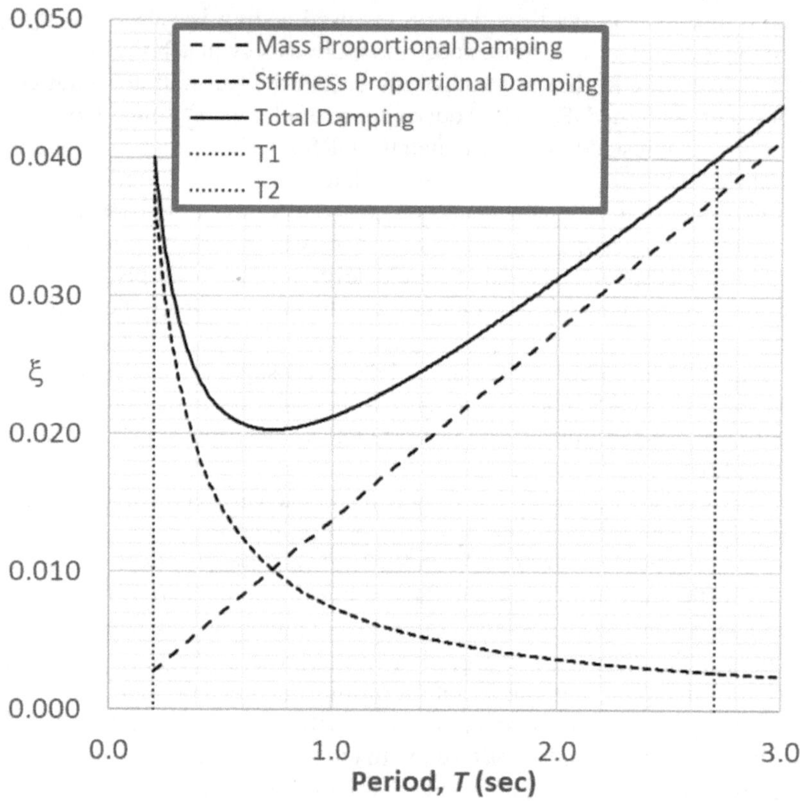

FIGURE 7.54 Rayleigh damping example.

Figure 7.54 shows a typical Rayleigh damping curve. The figure depicts damping, as a percent of critical damping, on the *y*-axis and natural period, *T*, on the *x*-axis. In this example, 4% damping at periods of 0.2 and 2.7 seconds has been specified. Calculations using Equations 7.68 and 7.69 show that the constants required to produce this result are $\alpha = 0.173329$ and $\beta = 0.002371$. Default values in software must not be universally accepted. Specified values other than the defaults for α and β are often required and it is up to the engineer to ensure proper damping specification.

Observe in Figure 7.54 that at any period between 0.20 and 2.70 seconds, the damping is less than that specified at the endpoints of the period range.

7.14.4 Nonlinear Response History Analysis

One issue in nonlinear response history analysis is the proper specification of damping. For structures expected to be stressed well beyond the elastic range, the typical 5% damping may be excessive. Some argue that an initial elastic component of damping should be equal to 0% for nonlinear dynamics. Further complicating the issue is the idea of initial-stiffness-based versus tangent-stiffness-based damping.

Recall that both Rayleigh damping and stiffness-proportional damping establish a damping matrix with at least one term proportional to the stiffness matrix. In LRHA the stiffness matrix is constant while in NLRHA the stiffness matrix changes throughout the analysis. With initial-stiffness-based damping, the damping matrix is computed from the initial stiffness matrix and does not change. With tangent-stiffness-based damping, not only the stiffness matrix but also the damping matrix are updated as the analysis progresses. Many would agree that tangent-stiffness-based damping is more accurate for NLRHA.

The choice of hysteretic model is important in nonlinear analysis.

7.14.5 Ground Motion Selection and Modification for Response History Analysis

Factors important in the selection and modification for ground motion records (or record pairs for three-dimensional structural analysis) include each of the following (NEHRP Consultants Joint Venture, 2011), with match to spectral shape identified as the single most important factor.

- match to spectral shape of the target for individual records
- moment magnitude
- tectonic regime (shallow crustal earthquakes, subduction earthquakes, etc.)
- subsurface conditions at the project site and recording station
- source-to-site distance as measured by the Joyner-Boor distance (R_{jb}), the epicentral distance (R_{epi}), or some other source-to-site metric

Given that match to target spectral shape is critical, some measure of the quality of the match is required. Two proposed measures include the DRMS and the mean-square-error, MSE. Equations 7.70 and 7.71 provide expressions for these. Equation 7.73 gives a means of calculating a scale factor, *f*, which minimizes the MSE over a given range of periods:

$$D_{\mathrm{RMS}} = \frac{1}{N}\sqrt{\sum_{i=1}^{N}\left(\frac{(\mathrm{SA_{GM}})_{\mathrm{Ti}}}{\mathrm{PGA_{GM}}} - \frac{(\mathrm{SA_{TAR}})_{\mathrm{Ti}}}{\mathrm{PGA_{TAR}}}\right)^2} \tag{7.70}$$

$$\mathrm{MSE} = \frac{\sum w(T_i)\cdot\left\{\ln\left[\mathrm{SA_{TARGET}}(T_i)\right] - \ln\left[f\cdot \mathrm{SA_{GM}}(T_i)\right]\right\}^2}{\sum w(T_i)} \tag{7.71}$$

$$\ln f = \frac{\sum\left[w(T_i)\cdot \ln\frac{\mathrm{SA_{TARGET}}(T_i)}{\mathrm{SA_{GM}}(T_i)}\right]}{\sum w(T_i)} \tag{7.72}$$

The MSE for a ground motion record (or record pair) may be computed both pre-scaled (using a scale factor $f = 1$ in Equation 7.71) and post-scaled (using the scale factor determined from Equation 7.72 in Equation 7.71). The post-scaled MSE is the appropriate measure of match to spectral shape when MSE criteria are adopted. The DRMS is independent of any scaling.

Ground motion selection and modification for response history analysis generally involves steps similar to the following, with deviations as required by the adopted design specification.

1. Develop a target response spectrum (TRS) for the project consistent with the governing code requirements. The TRS may or may not be identical to the design response spectrum (DRS).
2. Establish the seismic sources contributing significantly to the hazard at the project site. Determine the magnitude range, recording station site class, and tectonic regime(s) applicable to the project. Disaggregation of the seismic hazard may be useful for this purpose and may be accomplished using tools available at the United States Geological Survey website.
3. Collect a suite of candidate ground motion record pairs using the criteria established in Step 2. The size of the initial suite of candidates should be relatively large, perhaps 100 or more records.
4. Using one of the metrics available, rank the record pairs in order of decreasing match to the target response spectrum shape. The metric could be DRMS, post-scaled MSE, or some other parameter. Post-scaled MSE appears to be a choice which often leads to a successful process with a minimum number of iterations.
5. Compute the scale factor for each candidate record using, for example, Equation 7.72.
6. Select the required number of records to be used in the response history analysis based on advantageous MSE (a lower MSE represents a superior match to the spectral shape of the target), desired scale factor, and subject to limitation on the number of records permitted from any single event. Desirable scale factors are generally in the range of 0.50–2.00, but some research has suggested that very large factors are acceptable as well. Codes and specifications may have explicit limits on the number of records permissible from a single event in a design suite of ground motions. In the absence of such explicit limitations, a reasonable number may be 30% of the total from any single event.
7. Evaluate the suitability of the scaled record suite using the governing code criteria. Commonly encountered criteria may include maximum permissible deviation of the suite-mean spectra from the target spectrum, minimum required ratio of suite-mean to target spectra over the period range of interest, maximum permissible ratio of suite-mean to target spectra over the period range of interest, required variability of the suite (typically expressed as a target natural-logarithm-based standard deviation) over the period range of interest, pulse-type content of records, etc.
8. If the suite does not satisfy the required criteria, options include spectral matching (when permitted by the governing specification) and re-scaling of the suite, a return to Step 3 and construction of a new set of candidate records, and expanding the size of the candidates, among others.

Three means of ground motion record modification are among those available and considered as potentially acceptable methods among modern design codes include:

- amplitude scaling
- spectral matching in the time domain
- spectral matching in the frequency domain

Each of these is discussed in the subsequent sections.

7.14.5.1 Amplitude Scaling of Ground Motion

Amplitude scaling is typically the preferred method for ground motion modification. The accelerations at each time step in the acceleration are all multiplied by some factor, such as that given by Equation 7.72. The time scale is not adjusted in any fashion. Frequency content and pulse-type character of the ground motion are retained.

Equations 7.71 and 7.72 are particularly useful in scaling ground motion records and record pairs. Tools available for computing scale factors which minimize MSE include:

- the PEER Ground Motion Database (https://ngawest2.berkeley.edu/) for records from worldwide shallow crustal earthquakes
- the NGA Subduction Portal (Mazzoni, 2022) for records from worldwide subduction earthquakes
- SigmaSpectra (https://github.com/arkottke/sigmaspectra) for scaling to both a target response spectrum and a target variability (Kottke & Rathje, 2012)

Other tools are becoming available as the need for suitable ground motion modification has grown. SigmaSpectra may be particularly useful as future performance-based design criteria could potentially include target variability, in addition to target PSA, for ground motion suites.

7.14.5.2 Spectral Matching in the Time Domain

Wavelets may be strategically added to ground motion records to adjust the spectral character of the record.

SeismoMatch (SeismoSoft, 2010) adds wavelets to an accelerogram to create a new accelerogram whose PSA response spectrum matches, as closely as possible, the target spectrum. Modification is performed directly on the accelerogram in the time domain.

It is preferable to always examine the characteristics of the modified ground motion and ensure that no important characteristics have been lost in the matching process. For example, near-field sites often require that pulse-type accelerograms be included when response history analysis is used in the design. Spectral matching, whether time-domain-based or frequency-domain-based can alter or completely erase the pulse characteristics of a ground motion.

It is also necessary to examine the integrated ground velocity and displacement histories of a spectrally matched record for anomalous characteristics.

Figure 7.55 shows the original and modified ground motion records, integrated twice to examine the ground displacement history, for an example target spectrum

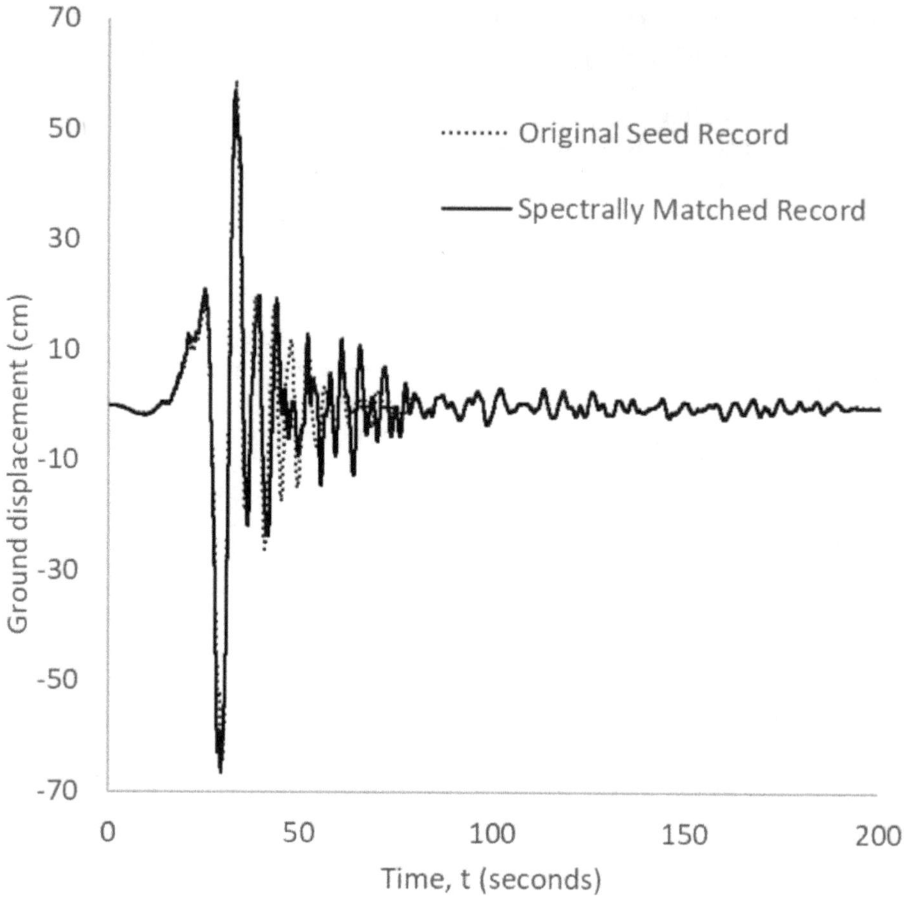

FIGURE 7.55 Original and spectrally matched ground motion records.

problem. The target displacement spectrum, the original accelerogram displacement response spectrum, and the spectrally matched accelerogram displacement response spectrum are shown in Figure 7.56. The matching for the depicted example was performed with SeismoMatch using a period range for the matching of 0.1–4.0 seconds. Clearly, while there is little observed difference in the original and spectrally matched ground motion histories, a significantly improved match to the target spectrum has been achieved through the process.

7.14.5.3 Spectral Matching in the Frequency Domain

SeismoArtif (SeismoSoft, 2012) has multiple capabilities, among which is spectral matching in the frequency domain. The Fourier spectrum for a record is first computed and compared to a Fourier spectrum generated from the target PSA spectrum. Adjustment to the ground motion Fourier spectrum is made to more closely match that target Fourier spectrum. The modified Fourier spectrum is then converted back to a new accelerogram.

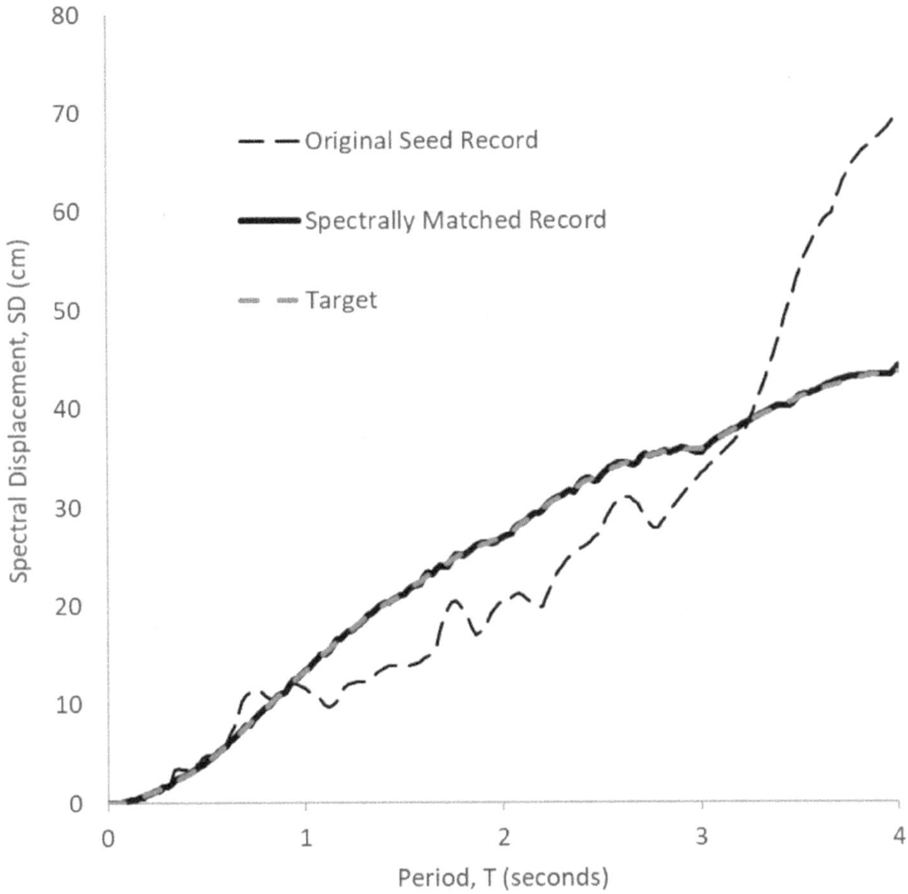

FIGURE 7.56 Response spectra – target, original, and spectrally matched.

RSPMATCH is another tool often used for spectral matching.

REQPY (Montejo, 2021) provides the capability to alter ground motion record pairs such that the maximum-direction (*RotD100*) spectrum of the record pair closely matches a target response spectrum. REQPY offers the ability, in fact, to match any *RotDnn*-based spectrum of ground motion record pairs to a target spectrum. Orientation-independent spectra in ASCE 7-22, for example, are maximum-direction spectra, also known as *RotD100* spectra.

Spectral matching procedures will at times produce unrealistic matched ground motions from a realistic seed record and should be used with care. Several matched records may have to be discarded before realistic accelerograms are produced. Criteria used to judge whether a generated accelerogram is realistic may include, among other factors, (1) preservation of frequency content between the original and matched records, (2) velocity or displacement drift and the end of the generated record, and (3) simple visual inspection of the generated velocity and displacement histories by integrating the matched accelerogram.

7.14.5.4 Synthetic and Artificial Accelerograms

In addition to the use of modified real, recorded ground motion records, methods exist for the generation of synthetic and artificial accelerograms based on either seismological theory or random signal generation and modification.

Synthetic accelerograms are generated from physics-based models of the faulting process. Tectonic regime (inter-plate, intra-plate, etc.), earthquake magnitude, subsurface conditions, and source-to-site distance are a few of the parameters that have been used to generate synthetic accelerograms in the literature.

Artificial accelerograms are generated from a random signal, which is subsequently modified in the frequency domain to improve match to a target response spectrum.

Refer to Section 7/7 for more information on synthetic and artificial accelerograms.

7.14.5.5 Example Problem: Ground Motion Selection and Modification

Using the hypothetical *RotD100*-based target response spectrum ordinates from the second column of Table 7.24, subduction ground motion records will be selected and modified using both spectral matching in the time domain and amplitude scaling. In addition to the target response spectrum, a target natural-logarithm-based standard

TABLE 7.24
Target Response Spectrum (TRS) for Selection Example

Period, *T*	*RotD100* DRS	*RotD100* to *GeoMean* Ratio	*GeoMean* TRS
0.001	0.453	1.000	0.453
0.010	0.547	1.200	0.456
0.020	0.549	1.200	0.458
0.030	0.577	1.200	0.481
0.050	0.695	1.200	0.579
0.075	0.883	1.200	0.736
0.100	1.052	1.200	0.877
0.150	1.234	1.200	1.028
0.200	1.318	1.200	1.098
0.250	1.344	1.203	1.117
0.300	1.333	1.206	1.105
0.400	1.232	1.213	1.015
0.500	1.113	1.219	0.913
0.750	0.836	1.234	0.678
1.000	0.646	1.250	0.517
1.500	0.414	1.253	0.331
2.000	0.298	1.256	0.237
3.000	0.199	1.261	0.158
4.000	0.156	1.267	0.123
5.000	0.139	1.272	0.109
7.500	0.115	1.286	0.089
10.000	0.109	1.300	0.084

deviation of 0.6 across the entire period range of interest will be adopted for this example. In practice, target spectra may be uniform hazard, conditional mean, or otherwise. Target response spectra (TRS) may or may not be identical to the design response spectra (DRS). Target standard deviation may or may not be required, depending on the design specification and codes to be enforced. Regardless, this example illustrates one method of ground motion selection and modification which may be useful in practice. It is critical to understand each of the following prior to beginning the process:

- the directional/component nature of the target response spectrum (geometric mean, *RotD100*, *RotD50*, etc.)
- the probabilistic basis of the target response spectrum (uniform hazard, conditional mean, risk-targeted, etc.)
- the characteristics which define the seismic sources contributing to the hazard at the project site (modal and mean magnitudes, site class, source-to-site distance, tectonic regime, etc.)

For this example, only subduction records from events having moment magnitude between 7.60 and 8.30 (7.95±0.35) were recorded at stations with an average shear wave velocity in the upper 30 m, V_{S30}, between 220 and 410 m/s (315%±30%). This is not to say that the ranges used for this example should always be applied in practice. But some rational means of setting acceptable ranges for the various parameters should be employed and must be consistent with the requirements of the governing codes. Note, as well, that this example is for a hypothetical project near a subduction zone.

The procedure for selection and modification in this example will assume that geometric mean-based scaling is employed. That is to say, the target response spectrum represents the geometric mean of two horizontal components. Hence, the geometric mean spectrum of each individual record pair is the component to be matched to the target. This represents a case for which the analyst has neither access to *RotD100* spectra of ground motion records nor tools to generate *RotD100* spectra for ground motion records. The first step for this example is then to convert the *RotD100*-based spectrum to the *GeoMean*-based target. The final column of Table 7.24 gives the spectral ordinates for the TRS.

The period range of interest for this example is 0.04–5.00 seconds, a relatively wide range by typical standards. A suite of no fewer than 11 record pairs is required. The required number of records varies among design specifications. A 12-record-pair suite will be developed in this example.

A suite of 70 candidate records was assembled based on the stated criteria. These records were initialed scaled and the post-scaled MSE was computed. The suite was then narrowed to a final size of 12 record pairs which were then spectrally matched to the TRS. ***SeismoMatch*** spectral matching in the time domain was used to add wavelets to the records. ***SigmaSpectra*** was then used to scale the spectrally matched records to the target TRS with the target natural-logarithm-based standard deviation of 0.6 across all periods of interest.

Table 7.25 provides the metadata for the 12-record-pair suite and Figure 7.57 displays the suite spectra with the suite geometric mean overlain with the target response spectrum. NGA Subduction records are identified by a record sequence

TABLE 7.25
Scale Factor Applied to Spectrally Matched Records

NGASub RSN	Region	M_W	V_{S30}, m/s	Earthquake	Station	f
3000419	Central America	7.66	392	Costa Rica	PPQR	0.6244
3001102	Central America	7.74	382	El Salvador	ST	0.5148
4003182	Japan	7.92	244	Ibaraki	HASAKI2	1.1801
4028572	Japan	8.29	255	Tokachi-oki	IKEDA	0.9277
4032566	Japan	8.29	274	Tokachi-oki	42111	2.4971
4032577	Japan	8.29	297	Tokachi-oki	47418	1.3338
4032611	Japan	7.74	295	Japan Sea	AKITA-S	0.7249
4040706	Japan	7.92	282	Ibaraki	KASHIMA	1.5341
6001252	South America	8.00	281	Pisco	ICA2	0.347
6001750	South America	7.98	270	Chile	LLCRF	1.0457
6001757	South America	7.98	345	Chile	LLST	0.8218
6002710	South America	7.82	260	Ecuador	APO1	1.7568

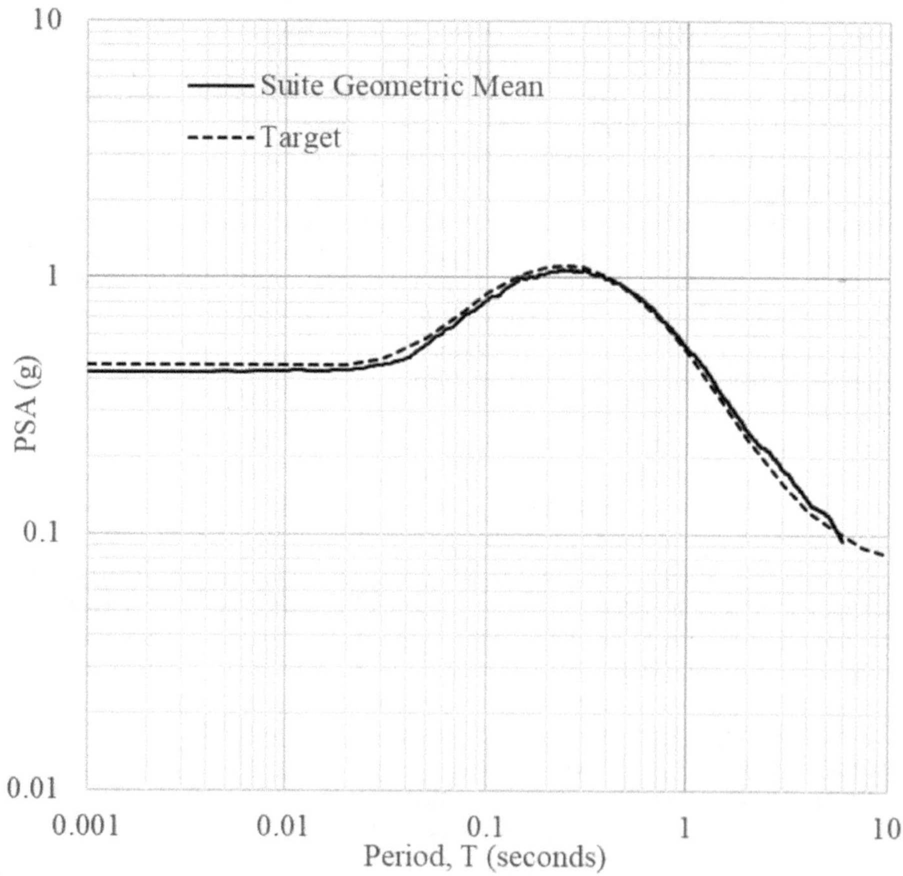

FIGURE 7.57 Suite mean vs. target response spectrum – selection example.

number (NGASub RSN), column 1 of Table 7.25. The final column of Table 7.25 is the scale factor applied to the spectrally matched records to achieve the target response spectrum and target variability in the suite.

The following summary would be helpful in establishing whether the resulting suite of records satisfies design criteria for a particular project. The summary is for the period range 0.04–5.00 seconds, the period range of interest for the example.

- maximum suite geomean-to-target ratio = 1.128
- minimum suite geomean-to-target ratio = 0.936
- average suite geomean-to-target ratio = 1.008
- minimum ln-based $\sigma = 0.514$
- maximum ln-based $\sigma = 0.795$
- average ln-based $\sigma = 0.697$

A note of caution is in order regarding spectral matching, whether in the time domain or in the frequency domain. It is always advisable to begin with candidate records which have been properly baseline adjusted and filtered, as evidenced by a zero velocity and displacement at the end of the records.

Amplitude scaling preserves the integrity of the records with regard to end-of-record drift in velocity and displacement. Spectral matching, on the other hand, may produce records with this end-of-record drift. The original accelerogram and the spectrally matched accelerogram may be difficult to distinguish when plotted. It is important to integrate the spectrally matched accelerogram to obtain the ground velocity history, and to integrate the ground velocity history to obtain the ground displacement history. Observe the end-of-record behavior and the overall nature of spectrally matched records. It may take many iterations to obtain acceptable spectrally matched records in practice.

Figures 7.58 (ground acceleration), 7.59 (ground velocity), and 7.60 (ground displacement) show both the original and the spectrally matched versions of NGASub RSN 4040706-090 (East/West component) used in this example. A slight displacement drift may be observed at the end of the record, which may or may not be acceptable depending on the criteria and the intended use. It is a good idea to eliminate any velocity or displacement at the end of the spectrally matched record. The effect will be more pronounced in many cases.

7.14.5.6 Example Problem: Damping Coefficient and Period Range Determination

Suppose a 3D model of a structure has been used to perform a modal analysis and the results are those summarized in Table 7.26. Vertical seismic loading is not applicable to this hypothetical problem.

- Determine the period range of interest to be used for linear response history analysis.
- Determine the period range of interest to be used for nonlinear response history analysis.

  Determine the Rayleigh damping coefficients to be used in linear response history analysis if the inherent damping is 3% of critical.

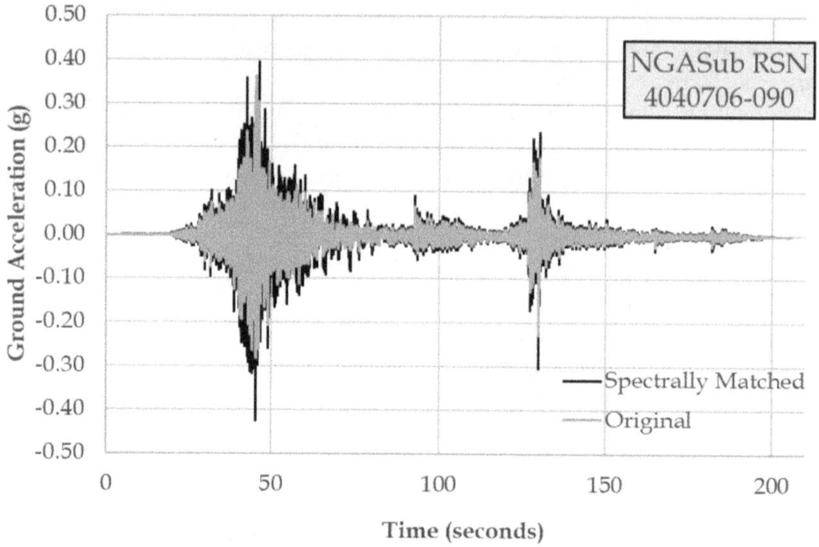

FIGURE 7.58 Original vs. spectrally matched RSN 4040706-090 (acceleration).

TABLE 7.26
Modal Analysis Results for Rayleigh Damping Example

Mode	*T*, seconds	Mass Participation, %			Cumulative Mass Participation, %		
		X: N/S	*Y*: E/W	*Z*: Vertical	*X*	*Y*	*Z*
1	1.62	0.0	43.6	0.0	0.0	43.6	0.0
2	1.25	56.9	0.0	0.0	56.9	43.6	0.0
3	0.86	0.0	39.5	2.6	56.9	83.1	2.6
4	0.66	3.3	0.0	0.0	60.2	83.1	2.6
5	0.41	0.0	12.9	0.1	60.2	96.0	2.7
6	0.40	0.1	0.0	17.6	60.3	96.0	20.3
7	0.32	0.0	1.0	0.0	60.3	97.0	20.3
8	0.26	14.7	1.5	0.0	75.0	98.5	20.3
9	0.22	0.5	0.2	0.2	75.5	98.7	20.5
10	0.21	0.0	0.1	16.6	75.5	98.8	37.1
11	0.20	10.2	0.1	0.0	85.7	98.9	37.1
12	0.18	0.0	0.0	0.0	85.7	98.9	37.1
13	0.17	0.0	0.0	0.6	85.7	98.9	37.6
14	0.15	6.6	0.0	0.0	92.3	98.9	37.6
15	0.14	0.0	0.4	0.0	92.3	99.3	37.6
16	0.12	2.5	0.0	0.0	94.8	99.3	37.7
17	0.11	0.0	0.0	0.0	94.8	99.3	37.7
18	0.10	0.0	0.0	48.0	94.8	99.3	85.7
19	0.08	0.0	0.0	0.0	94.8	99.3	85.7
20	0.08	5.0	0.0	6.9	99.8	99.3	92.6

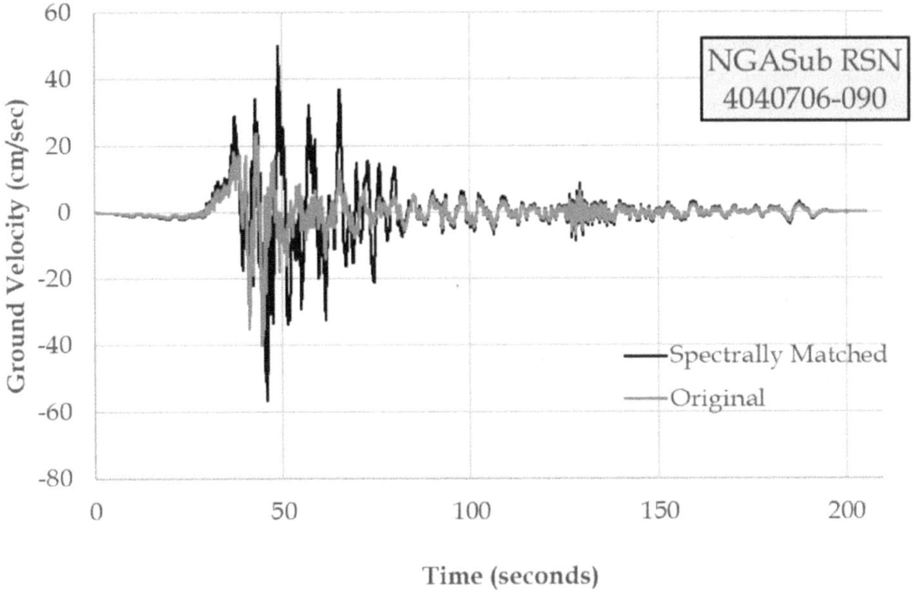

FIGURE 7.59 Original vs. spectrally matched RSN 4040706-090 (velocity).

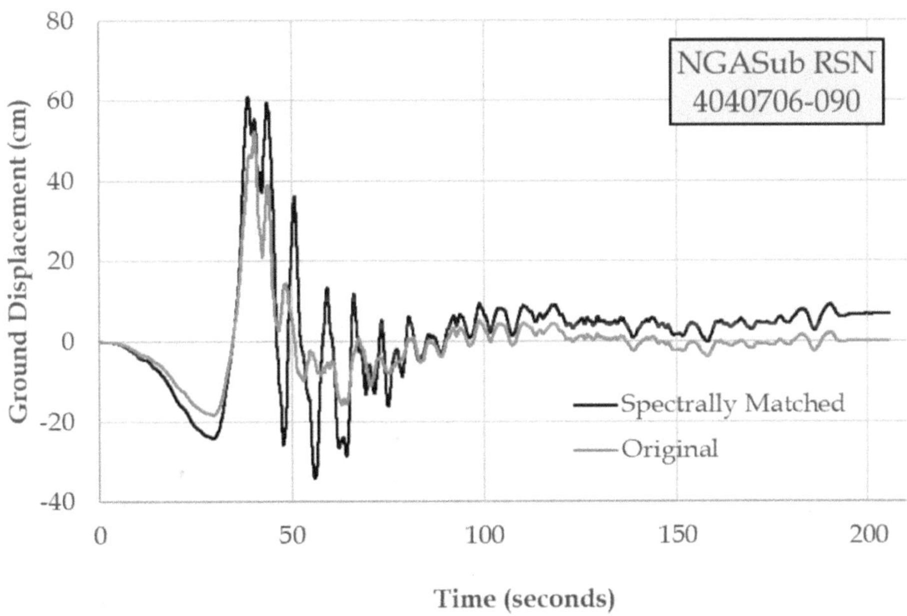

FIGURE 7.60 Original vs. spectrally matched RSN 4040706-090 (displacement).

Observe from Table 7.26 that the larger of the two fundamental periods is 1.62 seconds for the East-West direction. The smaller fundamental period is for the North-South direction and is equal to 1.25 seconds. Note that the period required to achieve at least 90% mass participation in both horizontal directions is 0.15 seconds.

For linear response history analysis, ASCE 7-22 requires a period range of interest equal to $[0.8T_{\text{lower}} - 1.2T_{\text{upper}}]$ (see Section 7.5 of this book). For this problem

- $T_{\text{lower}} = 0.15$ seconds
- $T_{\text{upper}} = 1.62$ seconds
- $T_{\text{range}} = [0.15 \times 0.8 - 1.62] = [0.12 - 1.62]$ seconds

For nonlinear response history analysis, ASCE 7-22 requires a lower bound on the period range of interest equal to the smaller of the period at which 90% mass participation has been achieved in both horizontal directions or 20% of the smaller fundamental period. The corresponding upper bound for nonlinear analysis is twice the larger fundamental period (see Section 7.5 of this book). For this problem

- $T_{\text{lower}} = \text{Min}(0.15, 0.20 \times 1.25) = 0.15$ seconds
- $T_{\text{upper}} = 2 \times 1.62 = 3.24$ seconds
- $T_{\text{range}} = [0.15\text{–}3.24]$ seconds

For the Rayleigh damping coefficients, ensure that no more than 3% of critical damping is present in any mode in the period range of interest for linear analysis. The appropriate coefficients may be computed using equations in Section 7.14.3 of this book:

$$\beta = \frac{2\xi_{\text{design}}}{\omega_1 + \omega_2} = \frac{T_1 T_2 \xi_{\text{design}}}{\pi(T_1 + T_2)} = \frac{0.12 \times 1.62 \times 0.03}{\pi(0.12 + 1.62)} = 0.001067$$

$$\alpha = \frac{2\omega_1\omega_2\xi_{\text{design}}}{\omega_1 + \omega_2} = \frac{4\pi\xi_{\text{design}}}{T_1 + T_2} = \frac{4\pi \times 0.03}{0.12 + 1.62} = 0.2167$$

Default values for Rayleigh damping coefficients typically exist in modern software. These defaults should never be blindly accepted but should be adjusted to the required values for a particular problem such as this one.

7.14.5.7 Example Problem: ELF, RSA, LRHA, and NLRHA

While more often than not, three-dimensional models of structures are required in analysis and design procedures, it may prove beneficial to examine a simple two-dimensional model and apply the principles from this and the preceding sections. Equivalent lateral force (ELF), modal response spectrum analysis (RSA), ground motion selection and modification, linear response history analysis (LRHA), and nonlinear response history analysis (NLRHA) procedures will be performed on the simple model shown in Figure 7.61.

The shear building model is shown in Figure 7.61. The first story height is 14 ft and the second story height is 12 ft.

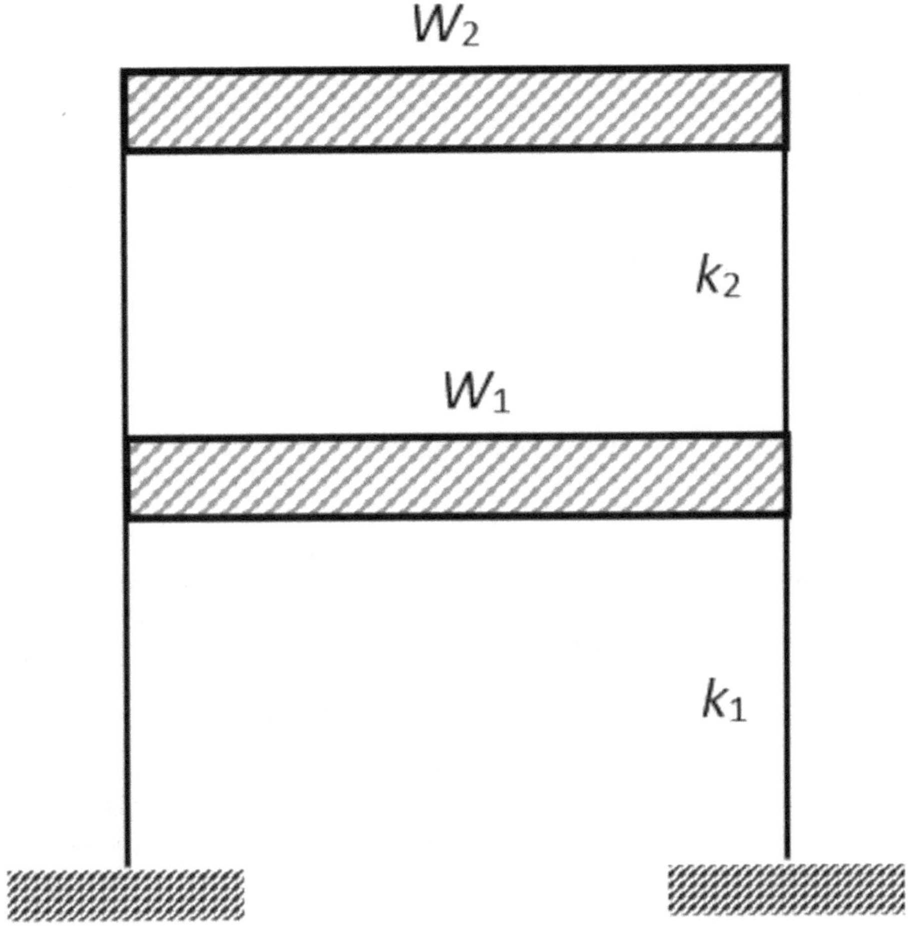

FIGURE 7.61 2D 2DOF shear building.

The following parameters have been established during preliminary design for the building. The shear wave velocity has been established not by direct measurement, but by correlation to standard penetration test blow counts. Hence, variability in subsurface conditions must be accounted for in the determination of the appropriate site classes:

$$W_1 = 1{,}000 \text{ kips} \quad k_1 = 523 \text{ kips/inch}$$

$$W_2 = 750 \text{ kips} \quad k_2 = 32 \text{ kips/inch}$$

$$V_{S30} = 822 \text{ ft/s} \quad \text{Risk Category} = \text{III}, \ I_e = 1.25$$

$$R = 3.25 \quad C_d = 3.25$$

TABLE 7.27
Design Ground Motion Parameters for 2DOF Building

Parameter	Site Class CD	Site Class D	Site Class DE
S_{MS}, g	1.399	1.213	0.949
S_{M1}, g	0.898	0.988	0.979
PGV, cm/s	93	83	69

Redundancy factor $\rho = 1.0$

For site classification, consider variability in correlations to blow count data:

$$V_{S30}/1.3 = 632 \text{ ft/s, Site Class} = \text{DE}$$

$$V_{S30} = 822 \text{ ft/s, Site Class} = \text{D}$$

$$V_{S30} \times 1.3 = 1,069 \text{ ft/s, Site Class} = \text{CD}$$

For this problem, ASCE 7-22 site classification with a 2-point design response spectrum will be used. For the hypothetical project site, Table 7.27 has been developed by assembling risk-targeted ground motion data and deaggregation of the seismic hazard at the project site.

Since variability in estimated shear wave velocity must be considered, take for design the largest value among the three site classes identified. Recall from previous material presented that design response spectra are two-thirds of the maximum considered values. Peak ground velocity (PGV) is not routinely included, but if the data are available, the parameter can be useful in ground motion selection and modification.

$$S_{MS} = 1.399 \quad S_{DS} = 0.933$$

$$S_{M1} = 0.988 \quad S_{D1} = 0.659$$

$$\text{PGV} = 93 \text{ cm/s} \quad \text{PGV} = 62 \text{ cm/s}$$

$$T_S = 0.706 \text{ seconds} \quad T_o = 0.141 \text{ seconds}$$

Hazard deaggregation at the project site indicates a modal moment magnitude of 7.55 and a corresponding source-to-site distance of 55 km. This will be used in ground motion selection for both linear and nonlinear response history analysis.

The dynamic characteristics of the shear building are summarized in Table 7.28. The large change in stiffness from the first story to the second story is intentional for this example, and illustrates the fact that for such irregularities, the ELF method may not be appropriate. The large change in stiffness also produces a model in which about 50% of the mass participates in each mode. This is unusual and it is common for the first mode to include a much larger percentage. In fact, 90% of the mass could be

TABLE 7.28
Dynamic Characteristics of 2DOF Model

Mode	Period, *T* (seconds)	Frequency (Hz)	% Mass	MCE PSA (g)	DRS PSA (g)
1	1.599	0.625	50	0.618	0.412
2	0.428	2.336	50	1.399	0.933

accounted for in the first mode for some models, in which case a single mode analysis could suffice for accuracy. Using only the fundamental mode for this particular problem would produce significant error, given the high percentage of mass in mode 2. The PSA values shown in Table 7.28 include those for both the maximum considered earthquake (MSE) and the design values (DRS) at two-thirds of the MCE.

For a period range of interest appropriate for linear response history analysis, take the lower end of the range equal to $0.80 \times 0.428 = 0.342$ seconds. Take the upper end of the range equal to 1.599 seconds:

$$T_{\text{range}} = [0.342 \text{ seconds}, 1.599 \text{ seconds}]$$

For a period range of interest appropriate for nonlinear response history analysis, take the lower end of the range equal to the lesser of (1) 0.428 seconds and (2) $0.2 \times 1.599 = 0.320$ seconds. Take the upper end of the range equal to $2.5 \times 1.599 = 4.000$ seconds:

$$T_{\text{range}} = [0.320 \text{ seconds}, 4.000 \text{ seconds}]$$

For this example, use the wider period range appropriate for nonlinear analysis since both linear and nonlinear analyses will be performed.

Ground motion records from earthquakes ranging in magnitude from 7.10 to 7.90 were considered. Recording stations located on Site Class C, D, and E subsurface profiles only were considered. Table 7.29 summarizes the selected records for a two-dimensional model analysis. The scale factors shown in the table are for the DRS spectrum, not the MCE spectrum. An increase in the factors by 50% is required to scale to the MCE. Three records are the minimum required for linear analysis. Eleven records is the minimum number required for nonlinear analysis. For this example, 14 records have been selected and scaled.

The average (geometric mean) PGV of the suite is 68 cm/s. While not a typical criterion for ground motion selection, this does enhance the validity of the selected suite since it is close to the value for design of 62 cm/s.

To confirm that the required match to the target response spectrum has been achieved, observe Figure 7.62. This shows the target DRS with the 14-record suite geometric mean spectrum. Also shown are the suite mean-to-target DRS ratio plot over the period range of interest, as well as the minimum suite mean-to target, average suite mean-to target, and maximum suite mean-to-target ratios. Over the period range of interest, the suite mean does not fall below 90% of the target DRS at any

TABLE 7.29
Ground Motion Records for 2DOF Model

Record	EQ	M_W	Station/Component	Scale	PGV, cm/s
1	Darfield	7.10	Resthaven/S88E	3.2639	86.87
2	El Mayor	7.20	Chihuahua/EW	3.1432	106.81
3	El Mayor	7.20	El Centro #3/EW	2.6361	78.42
4	El Mayor	7.20	Meloland/NS	2.6673	79.85
5	Landers	7.28	1000 Palms PO/N45E	2.4012	48.46
6	Taiwan	7.30	575/NS	2.9715	64.63
7	Taiwan	7.30	577/EW	2.8703	62.61
8	Chi-Chi	7.62	1163/EW	2.9231	51.74
9	Chi-Chi	7.62	1229/EW	2.5582	36.30
10	Chi-Chi	7.62	1339/NS	2.9999	22.04
11	Ecuador	7.80	AMNT/EW	2.2730	116.71
12	Synthetic	7.55	GM05/H2	1.4783	237.41
13	Wenchuan	7.90	Bixian/NS	2.5596	47.13
14	Wenchuan	7.90	Dayiyinping/EW	2.2487	73.63

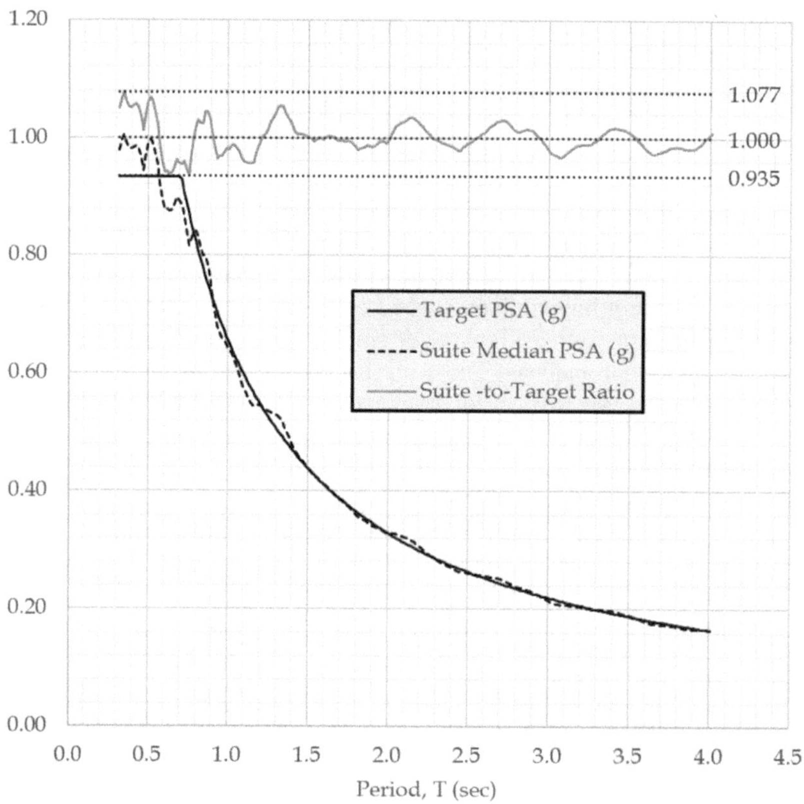

FIGURE 7.62 2DOF example ground motion suite spectra versus the DRS.

single period, the average suite mean-to target ratio is 1.000, and the maximum suite mean-to-target ratio is 1.077, less than a value of 1.3 required by some specifications.

While not a criterion for ground motion suites to date, it may become standard practice to require a target variability in the spectra of a ground motion suite. ASCE 7-22 specifies a target natural-logarithm-based standard deviation of 0.60 across all periods in the range of interest. While not explicitly stated as a requirement for ground motion selection and modification, it does seem informative to retain variability in the response to structures to earthquake ground motion. For this example, a natural-logarithm-based standard deviation target of 0.60 across the period range of interest has been used. Note that the average scale factor for the 14 ground motion records is 2.642. Applying this scale factor to each ground motion would produce similar results to those shown in Figure 7.62, but a different variability. In fact, applying any set of scale factors whose average is 2.642 would produce similar results as shown in Figure 7.62. SigmaSpectra (Kottke & Rathje, 2012) is a software package which maintains an average scale factor for a given suite, and adjusts the individual scale factors in an effort to achieve, not only match to a target response spectrum for the suite, but also match to a target natural-logarithm-based variability in the suite. Figure 7.63 depicts the target and suite variabilities for the selected records. While no specific deviation of the suite variability from the target variability has

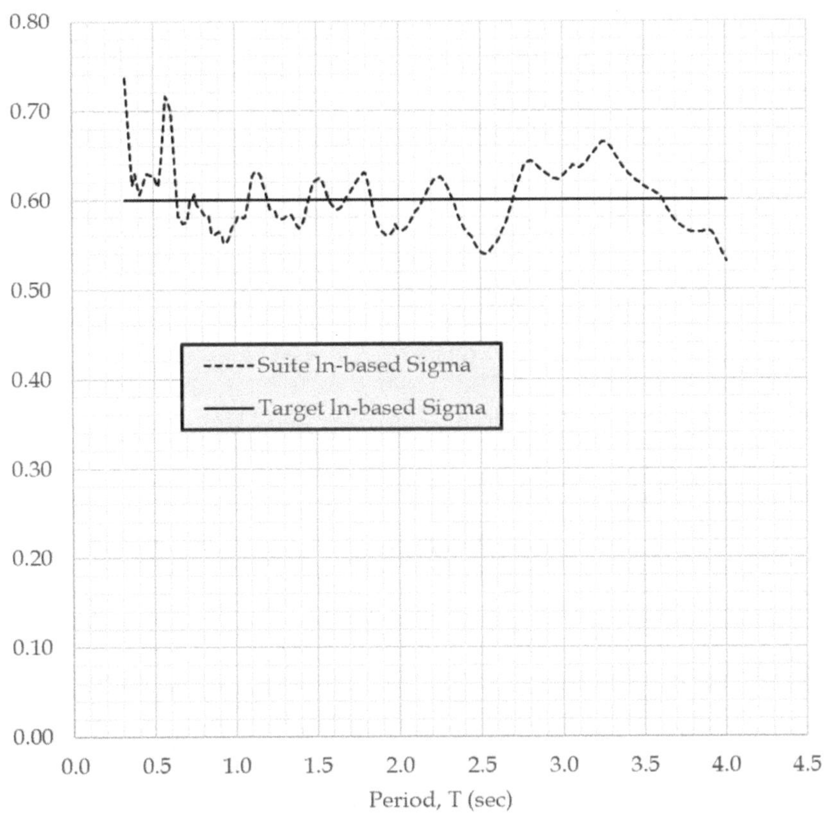

FIGURE 7.63 2DOF example suite variability versus target variability.

been specified, the result indicated in the Figure is taken as acceptable for this example. Future specifications for seismic design may also include criteria for permissible deviation in variability. This issue is not critical if governing specifications require design values corresponding to the mean or the maximum from a suite of analyses. However, future requirements could possibly include design values corresponding to the mean $+n\sigma$ response from a suite of analyses where n is some number of natural-logarithm-based standard deviations from the mean. For such a requirement, retaining variability in the ground motion would be critical.

For the equivalent lateral force analysis, response will first be computed based on the DRS with no reduction by the factor (R/I_e). The allowance for inelastic response is accounted for by adopting design force and shear values by this factor, but for comparison to RSA and RHA results, the factor will be excluded. For a fundamental period of 1.599 seconds, the PSA (and the seismic coefficient, C_s) without the application of reduction factors is:

$$C_s = \frac{0.659}{1.599} = 0.412$$

The total base shear is then:

$$V = 0.412 \times 1{,}750 = 721 \text{ kips}$$

For the vertical distribution of the base shear, the exponent k needs to be interpolated:

$$k = 1.00 + \frac{1.599 - 0.50}{2.50 - 0.50} = 1.55$$

$$\sum wh^k = 1{,}000 \times 14^{1.55} + 750 \times 26^{1.55} = 59{,}693 + 116{,}832 = 176{,}525$$

The ELF floor forces are:

$$F_1 = 721 \times (59{,}693/176{,}525) = 244 \text{ kips}$$

$$F_2 = 721 \times (116{,}832/176{,}525) = 477 \text{ kips}$$

The ELF story shears are:

$$V_1 = 721 \text{ kips}$$

$$V_2 = 477 \text{ kips}$$

The ELF story drift values are:

$$\Delta_1 = 721/523 = 1.38 \text{ inches}$$

$$\Delta_2 = 477/32 = 14.91 \text{ inches}$$

For the modal superposition response spectrum analysis, SeismoStruct has been used to perform the required operations. The modal analysis results have already been summarized in Table 7.28. Table 7.30 summarizes the RSA drift and story shear values from modal superposition with SRSS modal combination.

Note from Table 7.30 that the drifts need to be computed for each mode and combined using the square-root-sum-of-squares (SRSS) or the complete quadratic combination (CQC) rule. Drift values computed from the SRSS displacements will always be in error. For the problem at hand, using the SRSS displacements to compute the second story drift would give a value of 11.09 − 1.71 = 9.38 inches. This is less than the SRSS drift at the second story of 10.54 inches. The discrepancy would be larger for many similar problems.

Table 7.31 summarizes the story shears (V), floor displacements (δ), and story drifts (Δ) for the 14-record suite using a linear response history analysis with the records scaled to the DRS target with the factors shown in Table 7.29. The RHA was completed using SeismoStruct. The initial, elastic component of damping was taken

TABLE 7.30
2D 2DOF RSA Drift and Story Shear

DOF	V_1, kips	Δ_1, inches	δ_1, inches	V_2, kips	Δ_2, inches	δ_2, inches
Mode 1	360	0.69	0.69	333	10.40	11.09
Mode 2	816	1.56	1.56	54	1.69	0.13
SRSS	892	1.71	1.71	337	10.54	11.09

TABLE 7.31
Linear RHA Analysis Results: 2D 2DOF Model

Record	V_1, kips	Δ_1, inches	δ_1, inches	V_2, kips	Δ_2, inches	δ_2, inches
1	2,163	4.14	4.14	538	16.80	18.74
2	1,181	2.26	2.26	416	12.99	12.70
3	838	1.60	1.60	276	8.62	8.79
4	971	1.86	1.86	301	9.42	9.97
5	600	1.15	1.15	349	10.89	11.85
6	1,325	2.53	2.53	441	13.78	13.53
7	1,320	2.52	2.52	336	10.49	10.35
8	658	1.26	1.26	289	9.03	9.59
9	461	0.88	0.88	163	5.11	5.46
10	312	0.60	0.60	121	3.80	4.04
11	1,876	3.59	3.59	861	26.90	27.52
12	1,627	3.11	3.11	805	25.16	25.76
13	1,277	2.44	2.44	362	11.32	10.14
14	644	1.23	1.23	192	5.99	5.90
GeoMean	954	1.83	1.83	339	10.59	10.84
$\sigma_{\ln Y}$	0.5588	0.5580	0.5580	0.5512	0.5502	0.5467

TABLE 7.32
Comparison of Results: ELF vs. RSA vs. LRHA

Response	ELF	RSA	LRHA
V_1, kips	721	892	954
Δ_1, inches	1.38	1.71	1.83
δ_1, inches	1.38	1.71	1.83
V_2, kips	477	337	339
Δ_2, inches	14.91	10.54	10.59
δ_2, inches	16.29	11.09	10.84

equal to 5% of critical damping for the linear response history analysis. Observe from the table that the natural-logarithm-based standard deviation ($\sigma_{\ln Y}$) in structural response is similar to, and actually slightly less than, the target value of 0.60 used in ground motion scaling. This will not be the case for the nonlinear RHA to be performed and is an important consideration.

It is important to ensure that all results computed make sense to the engineer. Table 7.32 summarizes the results of the three analyses performed thus far – (1) the equivalent lateral force (ELF) method without reduction by the factor (R/I_e), (2) the modal superposition response spectrum analysis (RSA) without reduction, and (3) the linear response history analysis (LRHA).

ASCE 7-22 requires that displacements and drifts from a linear analysis be amplified by (C_d/I_e) to estimate inelastic displacements and drifts. The inelastic drift estimates for the project, based on the RSA linear results, are as follows.

$$(\Delta_1)_{\text{inel}} = 1.71 \times (3.25/1.25) = 4.45 \text{ inches}$$

$$(\Delta_2)_{\text{inel}} = 10.54 \times (3.25/1.25) = 27.40 \text{ inches}$$

Before a nonlinear response history analysis can be completed, the results of either the ELF or the RSA methods will be used to establish approximate yield levels for each story. ASCE 7-22 accounts for inelastic behavior by requiring the design forces to be those from either ELF or RSA reduced by the factor (R/I_e). For this example, NLRHA will incorporate a bilinear hysteretic behavior with a 0.5% post-yield stiffness ratio and zero damping for the initial elastic component of damping. Engineers are required to select appropriate hysteretic models depending on the inelastic mechanism. The required yield level for each story may thus be computed. Given that the RSA is presumably more accurate than the ELF method, results from the RSA will be used to establish yield levels for each story:

$$F_{y1} = \frac{892}{3.25/1.25} = 343 \text{ kips}$$

$$F_{y2} = \frac{337}{3.25/1.25} = 130 \text{ kips}$$

TABLE 7.33
Nonlinear RHA Analysis Results: 2D 2DOF Model

Record	V_1, kips	Δ_1, inches	V_2, kips	Δ_2, inches
1	399	22.12	134	27.73
2	375	12.96	137	46.35
3	356	5.65	134	28.95
4	364	8.65	132	16.24
5	347	2.19	133	22.69
6	367	9.79	134	30.40
7	358	6.59	135	33.16
8	348	2.40	131	10.79
9	346	1.96	131	12.87
10	345	1.30	131	7.66
11	381	15.34	138	56.44
12	448	40.77	140	66.06
13	355	5.21	133	21.15
14	353	4.56	132	19.40
GeoMean	364	6.37	134	24.12
$\sigma_{\ln Y}$	0.071	0.989	0.020	0.617

Observe that these yield levels produce yield drifts:

$$\Delta_{y1} = 343 / 523 = 0.66 \text{ inches}$$

$$\Delta_{y2} = 130/32 = 4.06 \text{ inches}$$

Table 7.33 summarizes results from the NLRHA for the same 14 ground motion records, but with scale factors increased by 50% since ASCE 7-22 requires the target response spectrum for NLRHA to be 1.5 times the DRS.

From Table 7.33 it becomes evident that the variability in drift from a nonlinear analysis significantly exceeds the variability in ground motion input, particularly for the first story (0.989 natural-logarithm-based standard deviation in drift versus 0.600 natural-logarithm-based standard deviation in the ground motion input).

Also, from Table 7.33 it is clear that the design values from a linear response history analysis (LRHA) do not provide the target level of response for this example with extreme irregularities. The amplified RSA drift for the first story is 4.45 inches, while the NLRHA-based first story mean drift is 6.37 inches. The amplified RSA drift for the second story is 27.40 inches, while the NLRHA-based first story mean drift is 24.12 inches. The first story drift was underestimated using linear RHA results amplified by the required factors. The second story drift was reasonably estimated using linear RHA results amplified by the required factors.

Figure 7.64 depicts the hysteretic response for the first story subjected to the scaled record #3. Figure 7.65 is the corresponding plot for the second story.

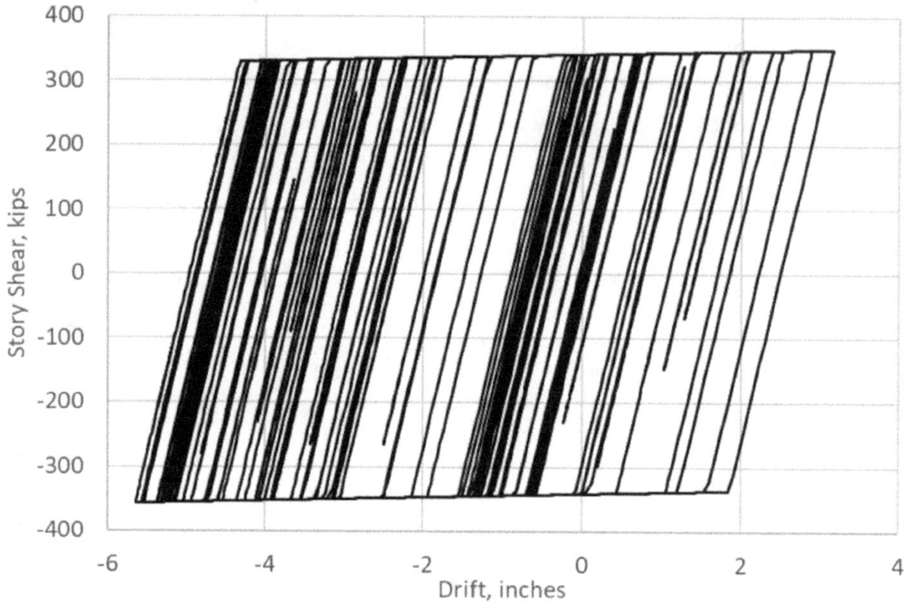

FIGURE 7.64 Hysteretic response for story 1: 2D 2DOF example.

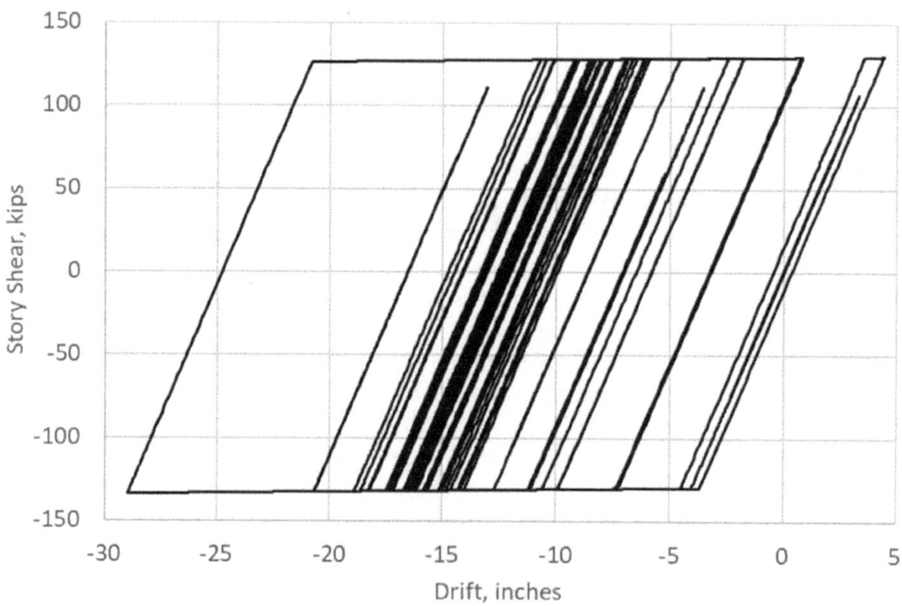

FIGURE 7.65 Hysteretic response for story 2: 2D 2DOF example.

FIGURE 7.66 SeismoStruct model for 2D 2DOF example.

The model for this example was composed of link-type elements and is depicted in Figure 7.66. The software used for all analyses is SeismoStruct.

While this simple example helps to illustrate the steps required for the various methods of analysis, the reader is referred to other excellent texts for a full treatment of analytical techniques used for seismic design. Some examples of these are Charney et al. (2020), Chopra (2016), Priestley et al. (2007), Priestley et al. (1996), and Clough and Penzien (1975). The help systems for software including SeismoStruct are also very useful for in-depth learning on analytical procedures.

7.14.6 Substitute Structure Method (SSM)

The substitute structure method (SSM) was introduced in Section 7.6.3 on inelastic response spectra. In fact, the SSM is the basis for the method presented for the generation of inelastic spectra directly from elastic spectra, given appropriate models for both (1) effective damping due to hysteretic response and (2) response modification due to increased damping.

7.14.7 Lateral Force-Resisting Systems for Buildings

Modern structural design for seismic effects often requires the consideration of expected, as opposed to minimum specified, material strengths. For structural steel, this is accounted for in AISC 341 through the use of yield and tensile strength factors, R_y and R_t, applied to minimum specified yield and tensile strengths, respectively. Table 7.34 provides a sampling of these factors for some commonly used steels.

Design methods in ASCE 7 require the specification of force-reduction factors (R-factors) and displacement amplification factors (C_d-factors) which vary depending on the lateral force-resisting system (LFRS) selected. Table 7.35 provides a sampling of values and height restrictions (h_{max}) for some commonly used LFRS for the different seismic design categories (SDC):

TABLE 7.34
Yield and Tensile Strength Factors from AISC 341

Material	R_y	R_t
A992 rolled shapes	1.1	1.1
A36 rolled shapes	1.5	1.2
A36 plates	1.3	1.2
A500 Grade B HSS	1.4	1.3
A500 Grade C HSS	1.3	1.2
A53 HSS	1.6	1.2
A1043 Grade 50 rolled shapes	1.2	1.1

TABLE 7.35
LFRS Design Parameters from ASCE 7

LFRS	R	C_d	SDC B h_{max}, ft	SDC C h_{max}, ft	SDC D h_{max}, ft	SDC E h_{max}, ft	SDC F h_{max}, ft
Special RC Shear Walls	5	5	NL	NL	160	160	100
Ordinary RC Shear Walls	4	4	NL	NL	NP	NP	NP
RC SMF	8	5.5	NL	NL	NL	NL	NL
RC IMF	5	4.5	NL	NL	NP	NP	NP
RC OMF	3	2.5	NL	NP	NP	NP	NP
Steel EBF	8	4	NL	NL	160	160	100
Steel SMF	8	5.5	NL	NL	NL	NL	NL
Steel IMF	4.5	4	NL	NL	35	NP	NP
Steel OMF	3.5	3	NL	NL	NP	NP	NP
Steel SCBF	6	5	NL	NL	160	160	100
Steel OCBF	3.25	3.25	NL	NL	35	35	NP
Special RM Shear Walls	5	3.5	NL	NL	160	160	100
Ordinary RM Shear Walls	2	1.75	NL	160	NP	NP	NP

- RC = Reinforced Concrete
- RM = Reinforced Masonry
- NL = No Limit
- NP = Not Permitted
- SMF = Special Moment Frame
- IMF = Intermediate Moment Frame
- OMF = Ordinary Moment Frame
- SCBF = Special Concentrically Braced Frame
- OCBF = Ordinary Concentrically Braced Frame
- EBF = Eccentrically Braced Frame

Also required is the seismic importance factor, I_e, which depends on the applicable risk category. The concept of risk category was introduced at the beginning of Section 7.14:

- Risk Category I, $I_e = 1.00$
- Risk Category II, $I_e = 1.00$
- Risk Category III, $I_e = 1.25$
- Risk Category IV, $I_e = 1.50$

Risk Category I includes structures, the failure of which, poses a low risk to human life. Examples include storage facilities and barns.

Risk Category II includes structures not classified as Risk Category I, III, or IV.

Risk Category III includes buildings and other structures, the failure of which could pose a substantial risk to human life.

Risk Category IV buildings and structures are designated as essential facilities and are intended to remain operational in the event of extreme ground shaking from earthquakes.

For additional guidance on the proper designation of risk category, refer to ASCE 7-22.

Lateral force-resisting systems (LFRS) in buildings are comprised of a fuse and capacity-protected elements. The fuse is the weakest, but also the most ductile, element in a properly designed LFRS in most modern design specifications. The fuse provides energy dissipation and limits the forces imparted to the capacity-protected elements.

Basic features of LFRS design for buildings is discussed in the remaining sections of this chapter. Seismic design for building systems is comprehensively discussed in the following standards.

- ANSI/AISC 341-22. *Seismic Provisions for Structural Steel Buildings*
- ANSI/AISC 358-22. *Prequalified Connections for Special and Intermediate Steel Moment Frames for Seismic Applications*
- ANSI/AISC 360-22. *Specification for Structural Steel Buildings*
- ASCE 7-22. *Minimum Design Loads and Associated Criteria for Buildings and Other Structures*

Section A.1 of ASCE 341 outlines the applicability of the various specifications and is given here verbatim:

> As specified in ASCE/SEI 7, Section 14.1.2.2.1, buildings with structural steel systems in seismic design categories B and C do not need to meet the requirements of these Provisions provided that they are designed in accordance with the AISC *Specification for Structural Steel Buildings* and the seismic design coefficients and factors of ASCE/SEI 7, Table 12.2-1, Item H. These Provisions do not apply in seismic design category A. ASCE/SEI 7 specifically exempts some systems from the requirements of these Provisions. Further discussion is provided in the commentary.

7.14.7.1 Moment Frames

Lateral resistance of moment frame systems is provided because of the moment connection of beams to columns. The energy dissipation mechanism – the fuse – in moment frame systems is plastic hinging near the ends of the beams which are part of the lateral force-resisting system. Properly designed moment frames are capable of relatively large lateral deformations before collapse.

Standard moment connections must be used based on design details and procedures in AISC 358-22, *Prequalified Connections for Special and Intermediate Steel Moment Frames.* Otherwise, special testing may be required to validate the performance of the connection. Moment connection types in AISC 358-22 include the following.

- Reduced beam section (RBS) connections
- Bolted unstiffened extended end plate connections (BUEEP)
- Bolted stiffened extended end plate connections (BSEEP)
- Bolted flange plate connections (BFP)
- Welded unreinforced flange – welded web connections (WUF-W)
- Cast bolted bracket connection (CBB)
- CONXTECH CONXL proprietary connections (CONXL)
- Sideplate connections
- Simpson Strong-Tie proprietary strong frame connections
- Double-Tee connections
- Slotted web connections (SW)
- Durafuse frames connections

The intended performance of many of the moment connection types is to move the plastic hinges in the beams away from the face of the column. Each connection type is limited in the conditions which its use is valid. Some of these limitations are summarized in Table 7.36. Refer to AISC 358-22 for the full limitations of each connection type.

One complication with the reduced beam section moment connection is the introduction of a non-prismatic beam section. This may be circumvented by applying the provision of Section 5.7 from AISC 358-22. This provision permits elastic drift calculations to be amplified by 10% to estimate the effect of the reduced beam section end regions of the beams.

TABLE 7.36
Prequalified Moment Connection Limitations

Connection	Beam Depth Maximum	Beam Weight Maximum	Column Depth Maximum	Clear Span-to-Depth Ratio
RBS	W44	408 plf	W40	7 or greater, SMF 5 or greater, IMF
BUEEP	AISC 358	AISC 358	W36	7 or greater, SMF 5 or greater, IMF
BSEEP	AISC 358	AISC 358	W36	7 or greater, SMF 5 or greater, IMF
BFP	W36	150 plf	W36	9 or greater, SMF 7 or greater, IMF
WUF-W	W36	150 plf	W36	7 or greater, SMF 5 or greater, IMF
CBB	W33	130 plf	W36	9 or greater
CONXL	W30	AISC 358	16″ Box	7 or greater, SMF 5 or greater, IMF
Sideplate	W44	529 plf	W44	See AISC 358-22
Simpson	W36	No limit	W36	No limit
Double-Tee	W24	55 plf	W36	9.0 or greater
SW	W36	400 plf	W36	6.4 or greater
Durafuse	W40	309 plf	W36	No limit

7.14.7.2 Concentrically Braced Frames

The fuse in a concentrically braced frame (CBF) consists of the braces, buckling in compression and yielding in tension. CBFs may be ordinary (OCBF) or special (SCBF). ConFigurations for concentrically braced frames include V-bracing, inverted V-bracing, X-bracing, and diagonal bracing. K-bracing conFigurations, in which the braces intersect at mid-height of a column, are typically not permitted. Modern design philosophy generally discourages inelastic behavior in columns which would result from the K-bracing conFiguration. Two-story X-bracing has also been proposed as an effective design conFiguration. One advantage of the two-story conFiguration is the potential reduction in seismically induced flexure in the beams. Multi-tiered frames, in which struts from column-to-column are located between floor levels, is also possible with CBF designs.

Figure 7.67 depicts a few of the various CBF conFigurations. The commentary to AISC 341-22 provides a more detailed discussion of viable conFigurations.

For V-bracing and inverted V-bracing OCBF design, AISC 341-22 requires that the required beam strength be determined, assuming no support from the braces for dead and live load.

For SCBF design, AISC 341-22 requires that at least 30%, but not more than 70%, of the horizontal force acting on a line of braces, be resisted by braces in tension.

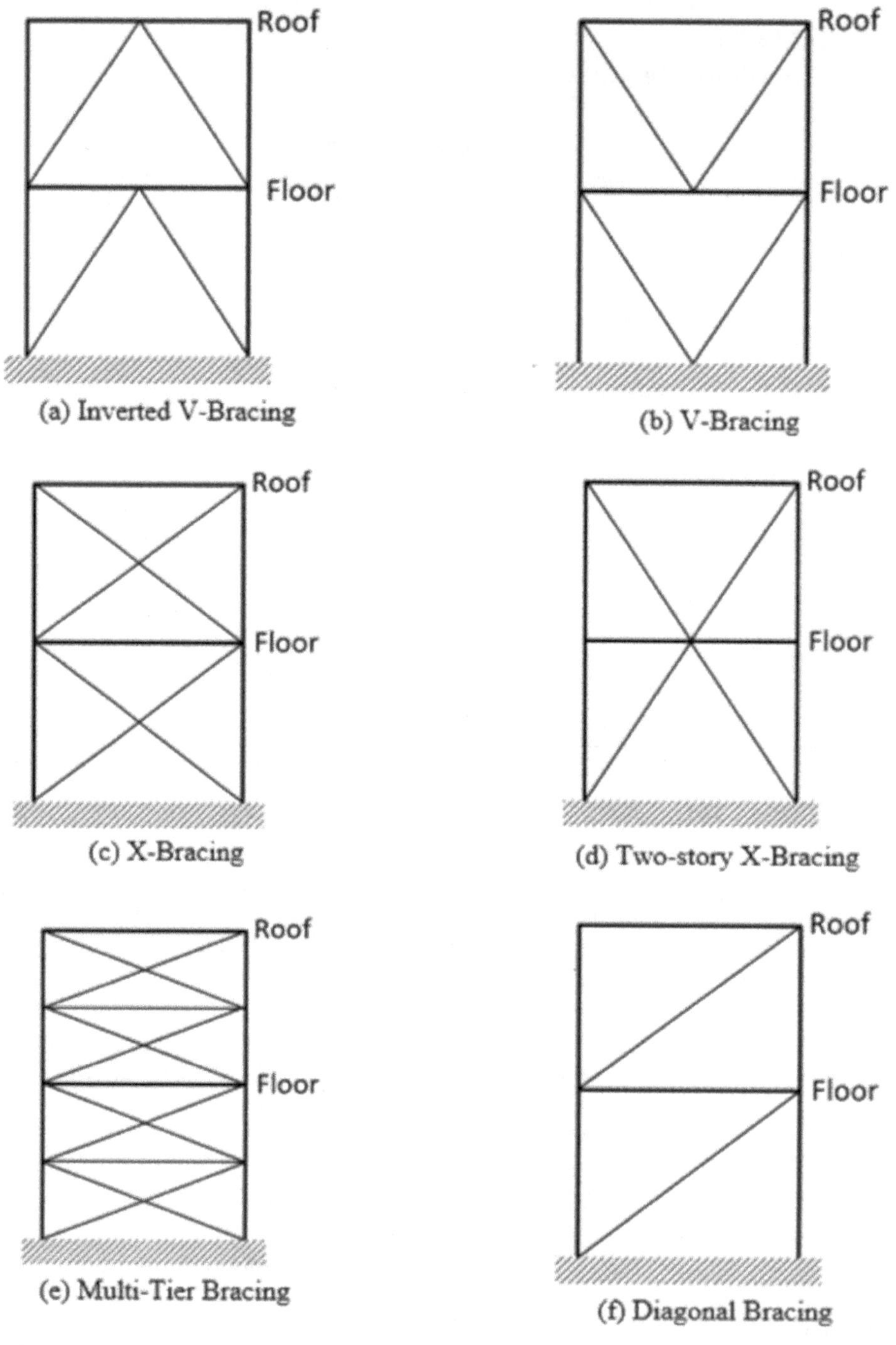

FIGURE 7.67 Concentrically braced frame Configurations.

This requires alternating the direction of diagonal braces in a given line. Columns, beams, and braces in SCBF construction must satisfy AISC 341-22, Section D1.1, cross-section requirements for highly ductile members.

Capacity protection of beams, columns, struts, and connections in SCBF construction must incorporate the capacity-limited seismic load effect, E_{cl}. This condition is to be taken as the larger of the following three cases:

a. Assume all braces carry a force equal to their expected strength in compression or in tension.
b. Assume all tension braces resist their expected strength and all compression braces resist their post-buckled expected strength.
c. For multi-tiered braced frames, an analysis of progressive yielding and buckling of braces from weakest to strongest.

For cases (a) and (b) AISC 341-22 defines the various strength computations as shown in Equations 7.73 through 7.75:

$$P_{Et} = R_y F_y A_g \tag{7.73}$$

$$P_{Ec} = \text{Min}\begin{cases} R_y F_y A_g \\ 1.14 F_{ne} A_g \end{cases} \tag{7.74}$$

$$P_{bc} = 0.3 P_{Ec} \tag{7.75}$$

- P_{Et} is the expected brace strength in tension.
- P_{Ec} is the expected brace strength in compression.
- P_{bc} is the post-buckling compressive strength of the brace.
- F_{ne} is the nominal resistance determined from AISC 360-16 Chapter E using the expected yield stress, $R_y F_y$, rather than the minimum specified yield stress, F_y, in the computations.

One somewhat problematic issue with the V-braced and inverted V-braced CBF is the design of the beams. As previously noted, AISC 341-22 requires that (1) no support from the braces is to be relied upon when designing the beams for dead and live loads, and (2) the braces essentially load the beams in the application of capacity-design principles for seismic effects. This often results in large beam sections. Even with the two-story CBF conFiguration, the issue is not eliminated.

7.14.7.3 Example Problem: Concentrically Braced Frame Calculations

Consider the two-story CBF depicted in Figure 7.68. The Figure shows the braces sizes required to resist the seismic forces in conjunction with live, snow, and dead loads with appropriate load factors ($U = 1.2D + 0.5L + 0.2S + 1.3E$). The beam must be capacity-protected by considering cases (a) and (b) in the discussion above. Figure 7.69 depicts the loading of the second-floor beam for the braces carrying their expected strength in both tension and compression (case a). Figure 7.70 depicts

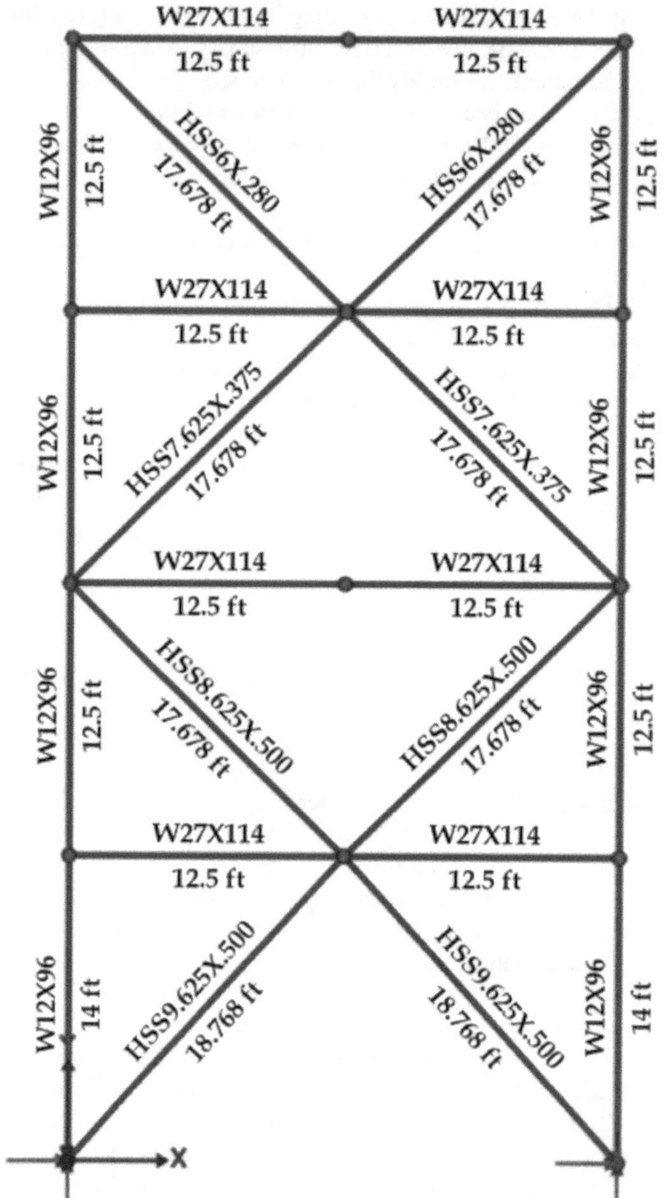

FIGURE 7.68 Example CBF Configurations.

the loading of the second-floor beam for the braces carrying expected strength in tension, post-buckled strength in compression (case b). Calculation of expected and post-buckled strengths have been completed as follows.

For the HSS braces made from A500 Grade B steel, $R_y = 1.4$ (see Table A3.2 in AISC 341-22) and $F_y = 42$ ksi. An effective length factor of K equal to 0.70 has

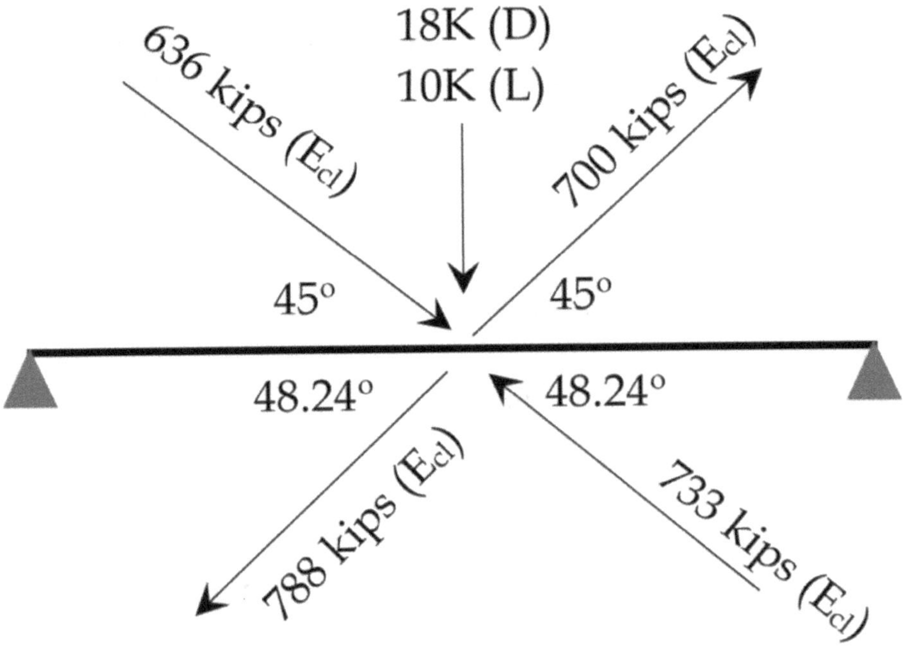

FIGURE 7.69 Example CBF – beams loads for case (a).

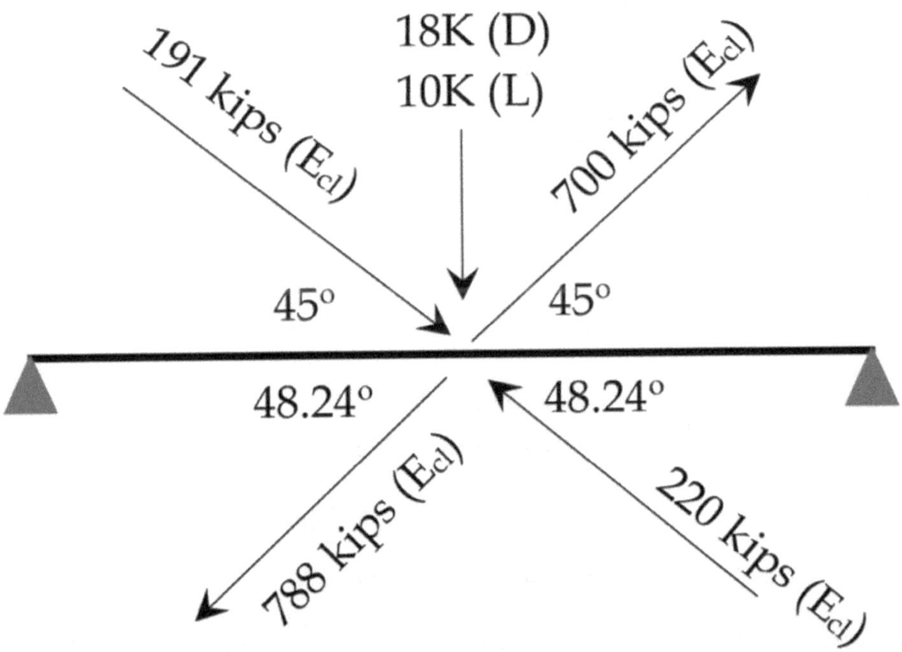

FIGURE 7.70 Example CBF – beam loads for case (b).

been assumed for these calculations. Careful consideration of connection details must be used in determining appropriate values for the effective length factor, K. The effective length factor is needed to determine the nominal compressive stress, F_{ne}, in accordance with AISC 360-22.

For the HSS 8.625×0.500 second story braces:

$$A = 11.9 \text{ inches}^2 \quad r = 2.89 \text{ inches}$$

$$F_e = \frac{\pi^2 E}{\left(L_c/r\right)^2} = \frac{\pi^2(29{,}000)}{\left(0.70 \times 212/2.89\right)^2} = 108.4 \text{ ksi}$$

$$\frac{R_y F_y}{F_e} = \frac{1.4 \times 42}{108.4} = 0.5424 < 2.25$$

$$F_{ne} = 0.658^{0.5424} \times (1.4 \times 42) = 46.9 \text{ ksi}$$

$$P_{Ec} = \text{Min}\left\{\begin{array}{l} 1.4 \times 42 \times 11.9 = 700 \text{ kips} \\ 1.14 \times 46.9 \times 11.9 = 636 \text{ kips} \end{array}\right. = 636 \text{ kips}$$

$$P_{Et} = 1.4 \times 42 \times 11.9 = 700 \text{ kips}$$

$$P_{bc} = 0.3 \times 636 = 191 \text{ kips}$$

For the HSS 9.625×0.500 first story braces:

$$A = 13.4 \text{ inches}^2 \quad r = 3.24 \text{ inches}$$

$$F_e = \frac{\pi^2 E}{(L_c/r)^2} = \frac{\pi^2(29{,}000)}{(0.70 \times 225/3.24)^2} = 121.1 \text{ ksi}$$

$$\frac{R_y F_y}{F_e} = \frac{1.4 \times 42}{121.1} = 0.4855 < 2.25$$

$$F_{ne} = 0.658^{0.4855} \times (1.4 \times 42) = 48.0 \text{ ksi}$$

$$P_{Ec} = \text{Min}\left\{\begin{array}{l} 1.4 \times 42 \times 13.4 = 788 \text{ kips} \\ 1.14 \times 48.0 \times 13.4 = 733 \text{ kips} \end{array}\right. = 733 \text{ kips}$$

$$P_{Et} = 1.4 \times 42 \times 13.4 = 788 \text{ kips}$$

$$P_{bc} = 0.3 \times 733 = 220 \text{ kips}$$

For case (a) shown in Figure 7.69, analysis of the simply supported W27×114 beam produces the following resultant horizontal (F_h) and vertical (F_v) forces at the center node from the capacity-limited loading, E_{cl}, alone.

$$F_h = (636 + 700)\cos 45° - (788 + 733)\cos 48.24° = -68.3 \text{ kips}$$

$$F_v = (-636 + 700)\sin 45° - (788 - 733)\sin 48.24° = +4.2 \text{ kips}$$

For case (b) shown in Figure 7.70, analysis of the simply supported W27 × 114 beam produces the following resultant horizontal (F_h) and vertical (F_v) forces at the center node from the capacity-limited loading, E_{cl}, alone.

$$F_h = (191 + 700)\cos 45° - (788 + 220)\cos 48.24° = -41.3 \text{ kips}$$

$$F_v = (-191 + 700)\sin 45° - (788 - 220)\sin 48.24° = -63.8 \text{ kips}$$

The E_{cl} forces add substantially to the forces from dead and live loads. Not also, that if this were an inverted V-brace, the upper brace forces would be absent. In this case, the downward load for case (b) would be:

$$F_v = (788 - 220)\sin 48.24° = -424 \text{ kips}$$

This simple calculation highlights the potential difficulties in capacity protection of the beams in V-brace and inverted V-brace CBF systems.

One proposed method of mitigating the potentially high economic impact on beam design for V-brace and inverted V-brace CBF systems is to permit some inelastic hinging at midspan of the beams (Tan et al., 2021). While promising, the method has yet to be codified in AISC 341.

7.14.7.4 Eccentrically Braced Frames

With an eccentrically braced frame (EBF), a short beam segment (the link) acts as the fuse through either flexural or shear (preferred) yielding. The braces connect to the beam eccentrically to form the link, which may be in the center of the beam or at the end of the beam. Figure 7.71 summarizes some common conFigurations for eccentrically braced frames.

During strong ground shaking, the link is subjected to large shear forces, as illustrated in Figure 7.72 for a four-story split-K EBF.

It is generally viewed as preferred practice to maintain an angle between the horizontal and the braces between 35° and 60°. Too large an angle may be detrimental to the stiffness of the system. Too small an angle complicates the connection and may introduce larger axial forces in the beam segments outside the links.

The links are first sized for the design seismic load, either through an equivalent lateral force procedure, a response spectrum analysis, or a response history analysis. Once the links have been sized to provide adequate strength and adequate displacement capacity, the braces, beam segments outside the links, and columns are designed using capacity-protection principles. This is accomplished by applying an overstrength loading, typically determined by incorporating expected yield strengths and factors to account for possible strain hardening in the links.

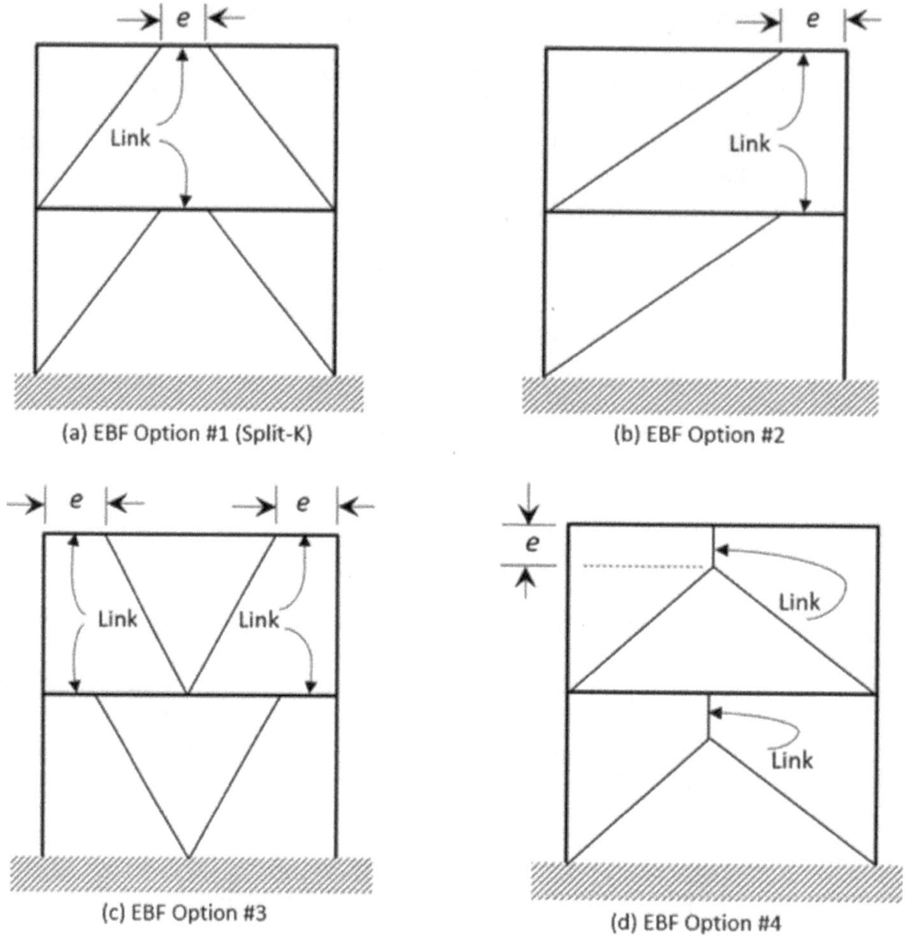

FIGURE 7.71 Eccentrically braced frame Configurations.

Figure 7.72 shows an EBF with the link located in the middle of the beam. This is referred to as a split-K EBF. The mechanics of a split-K EBF behavior are relatively simple, at least for preliminary design purposes using an equivalent lateral force design procedure. First, for a given lateral force distribution, the shear in the links may be estimated using Equation 7.76. Definitions for the parameters in Equation 7.76 are: h is the height of the story below the link, L is the bay width, V_{story} is the sum of the floor forces at and above the link under consideration. For the split-K EBF, the plastic displacement capacity, Δ_{pn}, is given by Equation 7.77, where γ_p is the plastic rotation capacity of the link and e is the link length (the length of the central beam segment bounded by the braces). The total drift demand consists of an elastic component and a plastic component. The plastic component may be estimated using Equation 7.78, where Δ_x is the drift from a structural analysis (for loads reduced by the force-reduction factor, R, from ASCE 7) and C_d is the displacement amplification factor from ASCE 7. For the EBF system, R is equal to 8 and C_d is equal

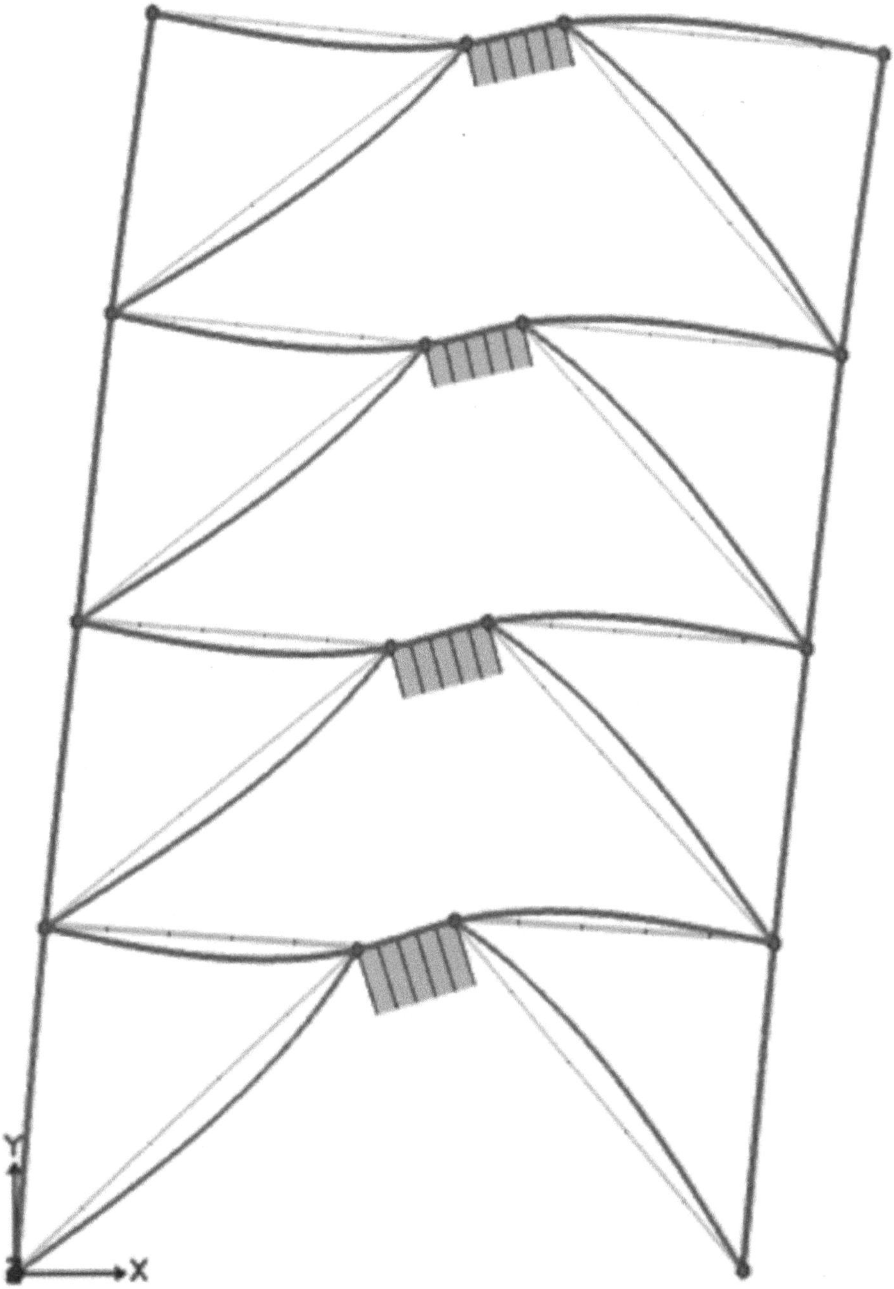

FIGURE 7.72 Link shears in an eccentrically braced frame.

to 4 in ASCE 7-22. The plastic drift demand must be less than or equal to the plastic displacement capacity, or the link must be resized:

$$V_{\text{Link}} \cong \frac{h}{L} V_{\text{story}} \tag{7.76}$$

$$\Delta_{\text{Pn}} = \frac{h}{L} \gamma_P e \tag{7.77}$$

$$\Delta_{\text{Pu}} = (C_d - 1)\Delta_x \leq \Delta_{\text{Pn}} \tag{7.78}$$

For plastic rotation capacity, the link must be classified as either a shear link, a flexural link, or an intermediate link, using criteria from AISC 341 (AISC, 2016). Equation 7.79 provides the means of classifying links. M_p is the plastic bending moment of the link and V_p is the plastic shear of the link. M_p and V_p are given in Equations 7.80 and 7.81, respectively. The equations are presented here in LRFD format, so the factored axial load, P_u, is the appropriate value. For the split-K EBF, P_u is typically small, and often less than $0.15F_yA_g$. This is not necessarily so for other EBF conFigurations, however.

The shear demand on a link from Equation 7.76 must be less than or equal to the reliable shear capacity, ϕV_n, given by Equation 7.82. Otherwise, the link must be resized. Only I-shaped and built-up box-shaped (rolled HSS shapes are not permitted) links are currently permitted in AISC 341.

The design of EBF systems is covered in detail in AISC 341 and other conFiguration options are shown there. For details not covered in this text, refer to AISC 341:

$$\begin{cases} e \leq 1.6\dfrac{M_P}{V_p} \rightarrow \text{Shear Link} \rightarrow \gamma_P = 0.08 \\ e \geq 2.6\dfrac{M_P}{V_p} \rightarrow \text{Flexural Link} \rightarrow \gamma_P = 0.02 \\ 1.6\dfrac{M_P}{V_p} < e < 2.6\dfrac{M_P}{V_p} \rightarrow \text{Intermediate Link} \rightarrow \text{Interpolate for } \gamma_P \end{cases} \tag{7.79}$$

$$M_P = \begin{cases} F_y Z & \text{for } \dfrac{P_u}{F_y A_g} \leq 0.15 \\ F_y Z\left(\dfrac{1 - P_u/(F_y A_g)}{0.85}\right) & \text{for } \dfrac{P_u}{F_y A_g} > 0.15 \end{cases} \tag{7.80}$$

$$V_p = \begin{cases} 0.6F_y A_{lw} & \text{for } \dfrac{P_u}{F_y A_g} \leq 0.15 \\ 0.6F_y A_{lw}\sqrt{1 - \left(\dfrac{P_u}{F_y A_g}\right)^2} & \text{for } \dfrac{P_u}{F_y A_g} > 0.15 \end{cases} \tag{7.81}$$

$$\phi V_n = 0.90 \times \text{Min} \begin{cases} V_p \text{ (shear yielding)} \\ 2M_p/e \text{ (flexural yielding)} \end{cases} \tag{7.82}$$

$$A_{\text{lw}} = \begin{cases} (d - 2t_f)t_w & \text{for I-shaped link sections} \\ 2(d - 2t_f)t_w & \text{for Box-shaped link sections} \end{cases} \tag{7.83}$$

AISC 341 includes ductility criteria for cross-sections. For EBF systems, the links must qualify as highly ductile, the braces must qualify as moderately ductile, and the columns must qualify as highly ductile. It is possible to use different cross-sections for the link and the beam outside the link. In such cases, the beam outside the link must qualify as moderately ductile. Chapter D of AISC 341 provides the means of qualifying cross-section elements. For I-shaped links, the web slenderness, h/t_w, and the flange slenderness, b/t, are the defining parameters of the cross-section which must be limited to values from the standard. The specification is freely available at AISC.org. The following example illustrates such calculations among others for eccentrically braced frames.

7.14.7.5 Example Problem: Eccentically Braced Frame Calculations

Figures 7.73 and 7.74 show a four-story EBF along with the proposed member cross-sections, frame geometry, and equivalent seismic loading reduced by the R-factor.

The second-floor displacement under the given loading is 0.36 inches.

Consider the second-floor link, a W10×77 section made from A992 steel ($F_y = 50$ ksi). Determine V_{story} for estimating the link shear:

$$V_{\text{story}} = 2(49.5 + 43 + 25.5 + 10.5) = 257 \text{ kips}$$

$$V_{\text{link}} \cong \frac{h}{L} V_{\text{story}} = \frac{14}{30} \times 257 = 120 \text{ kips}$$

Computer-aided structural analysis gives a link shear equal to 120.2 kips. The simple approximate expression gives a very good estimate for the link shear, as will typically be the case for split-K EBF systems.

Next, the properties of the W10×77 section are summarized.

$A = 22.7 \text{ inches}^2$	$Z_x = 97.6 \text{ inches}^3$	$d = 10.6$ inches	$b_f = 10.2$ inches
$t_w = 0.53$ inches	$t_f = 0.87$ inches	$h/t_w = 14.8$	$b/t = 5.86$

The properties were obtained from the 16th edition of the AISC Steel Construction Manual. If the Manual is not available, AISC provides an Excel file with the properties of all sections in the Manual. The Excel database is available at AISC.org. https://www.aisc.org/publications/steel-construction-manual-resources/16th-ed-steel-construction-manual/aisc-shapes-database-v16.0/

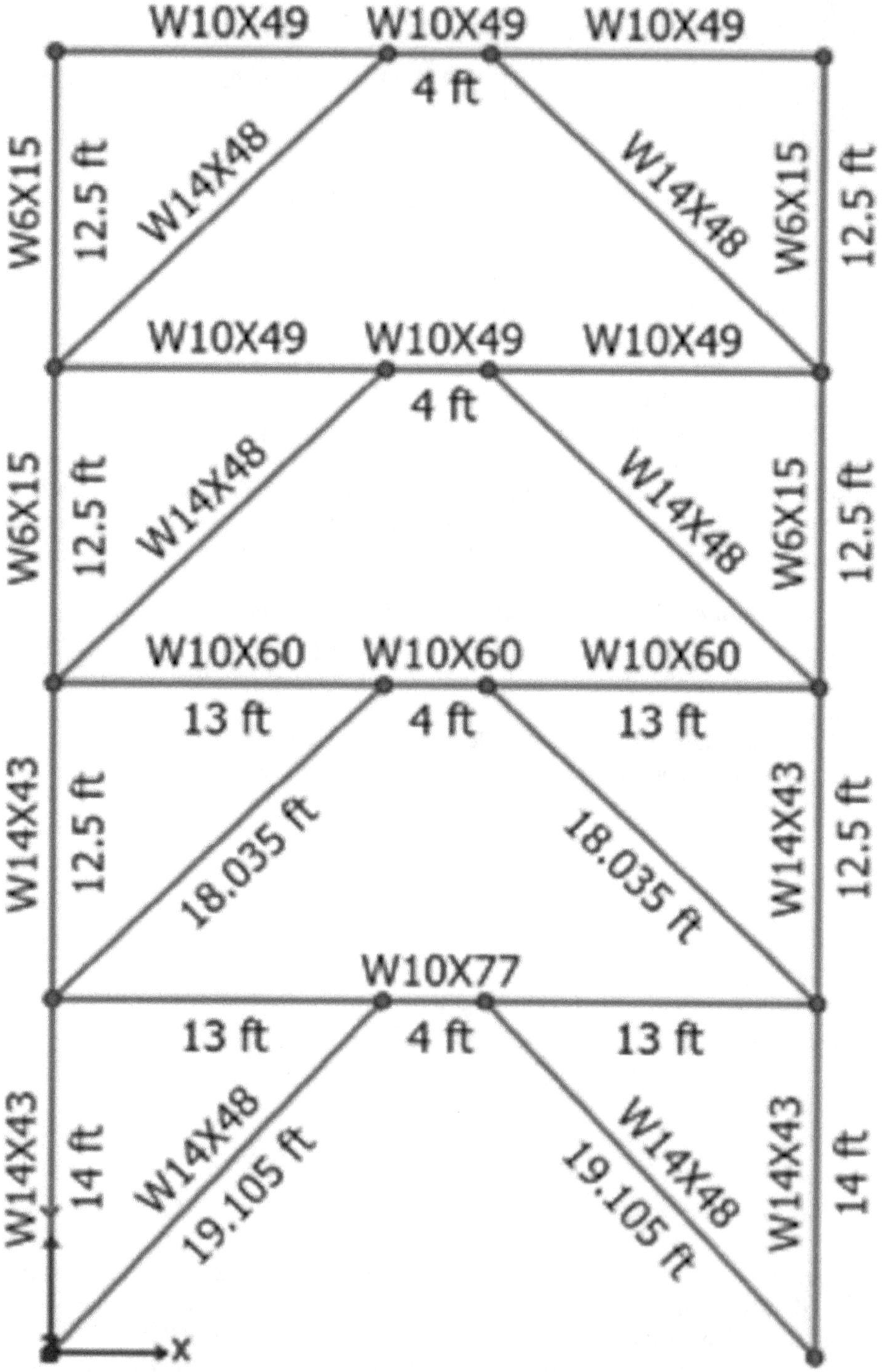

FIGURE 7.73 EBF example frame geometry.

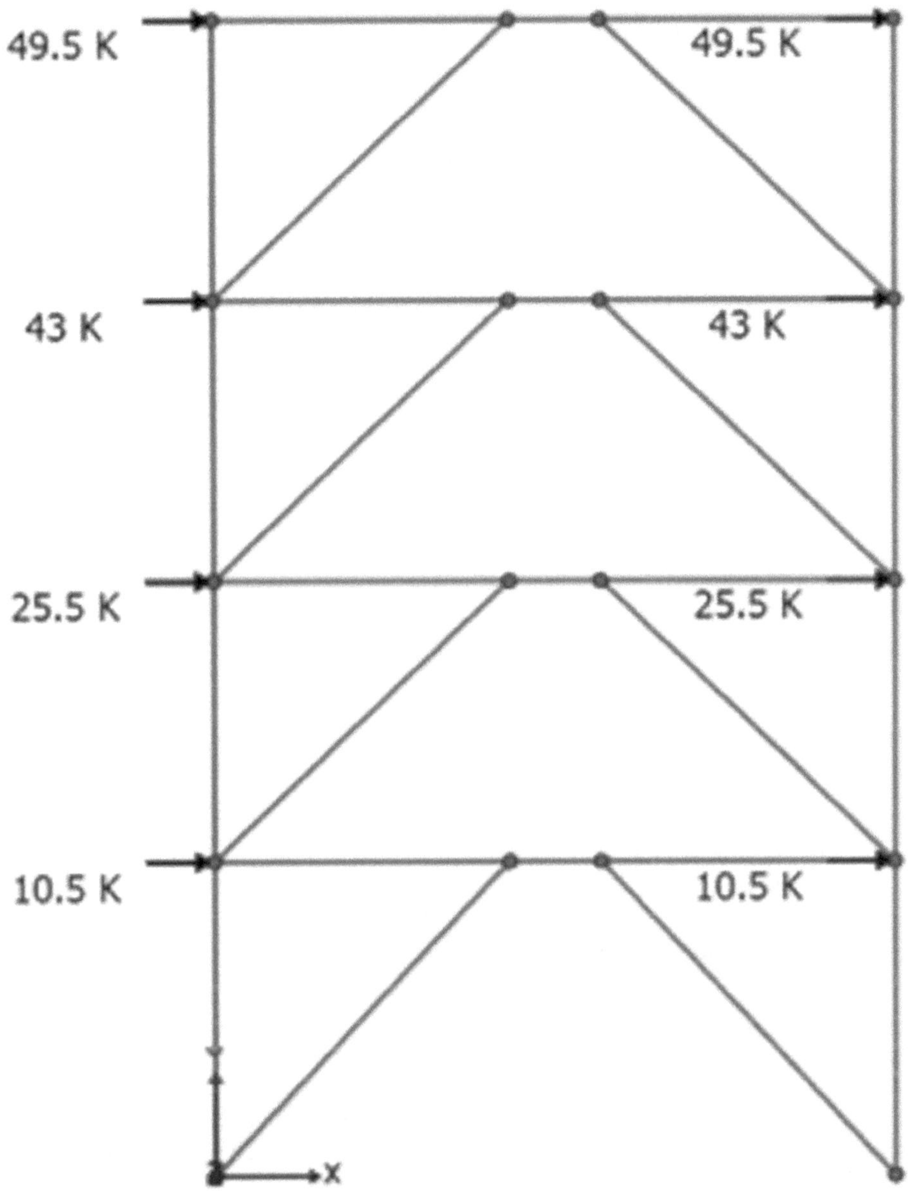

FIGURE 7.74 EBF example frame seismic loading.

It can be shown, using the given properties and the criteria from AISC 341, that the link section qualifies as highly ductile. Since the axial load in the link is approximately zero, the coefficient C_a is approximately zero:

$$\text{Flange}\frac{b}{t} = 5.86 < 0.32\sqrt{\frac{E}{R_y F_y}} = 0.32\sqrt{\frac{29{,}000}{1.1 \times 50}} = 7.35$$

$$\text{Web } \frac{h}{t_w} = 14.8 < 2.57\sqrt{\frac{E}{R_y F_y}}(1 - 1.04C_a) = 2.57\sqrt{\frac{29{,}000}{1.1 \times 50}} = 59.0$$

Compute the plastic moment and the plastic shear for the W10×77 link:

$$M_p = F_y \times Z_x = 50 \times 97.6 = 4{,}880 \text{ inch-kips}$$

$$V_p = 0.6 F_y A_{\text{lw}} = 0.6 \times 50 \times (10.6 - 2 \times 0.87) \times 0.53 = 141 \text{ kips}$$

The link length, $e = 4$ ft $= 48$ inches. Determine if the link qualifies as a shear link:

$$e = 48 \text{ inches} \leq 1.6 \frac{4{,}880}{141} = 55.4 \text{ inches} \rightarrow \text{Shear Link, } \gamma_p = 0.08 \text{ radians}$$

Determine the plastic drift capacity:

$$\Delta_{\text{Pn}} = \frac{14}{30} \times 48 \times 0.08 = 1.79 \text{ inches}$$

Determine the plastic drift demand:

$$\Delta_{\text{Pu}} = (C_d - 1)\Delta_x = (4 - 1)0.36 = 1.08 \text{ inches} \leq \Delta_{\text{Pn}} = 1.79 \text{ inches, OK}$$

Determine the shear resistance for the link and compare it to the shear demand on the link:

$$V_n = \text{Min} \begin{cases} V_p = 141 \text{ kips} \\ 2M_p/e = 2 \times 4{,}880/48 = 203 \text{ kips} \end{cases}$$

$$\Rightarrow \phi V_n = 0.90 \times 141 = 127 \text{ kips} > V_{\text{Link}} = 120 \text{ kips, OK}$$

The second floor W10×77 A992 link is satisfactory for both displacement capacity and strength. For capacity protection of the braces and beam and columns, AISC 341 requires that the link shear forces be set equal to the adjusted link shear strength given by Equation 7.84. With all link shears set to the adjusted link strength, the capacity-limited story shear at each level can be back-calculated from Equation 7.76:

$$(V_{\text{link}})_{\text{cl}} = \begin{cases} 1.25 R_y V_n & \text{for I-shaped links} \\ 1.40 R_y V_n & \text{for box-shaped links} \end{cases} \tag{7.84}$$

The adjusted link shears are first computed for each level of the frame:

Second floor: W10×77

$$(V2_{\text{link}})_{\text{cl}} = 1.25 \times 1.1 \times 141 = 194 \text{ kips}$$

Third floor: W10×60

$$M_p = F_y \times Z_x = 50 \times 74.6 = 3{,}730 \text{ inch-kips}$$

$$V_p = 0.6F_yA_{\text{lw}} = 0.6 \times 50 \times (10.2 - 2 \times 0.68) \times 0.42 = 111 \text{ kips}$$

$$V_n = \text{Min}\begin{cases} V_p = 111 \text{ kips} \\ 2M_p/e = 2 \times 3{,}730/48 = 155 \text{ kips}) \end{cases}$$

$$(V3_{\text{link}})_{\text{cl}} = 1.25 \times 1.1 \times 111 = 153 \text{ kips}$$

Fourth floor: W10×49

$$M_p = F_y \times Z_x = 50 \times 60.4 = 3{,}020 \text{ inch-kips}$$

$$V_p = 0.6F_yA_{\text{lw}} = 0.6 \times 50 \times (10.0 - 2 \times 0.56) \times 0.34 = 91 \text{ kips}$$

$$V_n = \text{Min}\begin{cases} V_p = 91 \text{ kips} \\ 2M_p/e = 2 \times 3{,}020/48 = 126 \text{ kips}) \end{cases}$$

$$(V4_{\text{link}})_{\text{cl}} = 1.25 \times 1.1 \times 91 = 125 \text{ kips}$$

Roof: W10×49

$$M_p = F_y \times Z_x = 50 \times 60.4 = 3{,}020 \text{ inch-kips}$$

$$V_p = 0.6F_yA_{\text{lw}} = 0.6 \times 50 \times (10.0 - 2 \times 0.56) \times 0.34 = 91 \text{ kips}$$

$$V_n = \text{Min}\begin{cases} V_p = 91 \text{ kips} \\ 2M_p/e = 2 \times 3{,}020/48 = 126 \text{ kips}) \end{cases}$$

$$(V5_{\text{link}})_{\text{cl}} = 1.25 \times 1.1 \times 91 = 125 \text{ kips}$$

Use the capacity-limited link shears to back-calculate the story shear at each level for the capacity-limited condition.

Fourth story:

$$V_{\text{story}} \cong \frac{L}{h}(V_{\text{link}})_{\text{cl}} = \frac{30}{12.5} \times 125 = 300 \text{ kips}$$

$$F4 = 274 \text{ kips}$$

Third story:

$$V_{\text{story}} \cong \frac{L}{h}(V_{\text{link}})_{\text{cl}} = \frac{30}{12.5} \times 125 = 300 \text{ kips}$$

$$F3 = F4 - 300 \text{ kips} = 0$$

Second story:

$$V_{\text{story}} \cong \frac{L}{h}(V_{\text{link}})_{\text{cl}} = \frac{30}{12.5} \times 153 = 367 \text{ kips}$$

$$F2 = 367 - 300 = 67 \text{ kips}$$

First story:

$$V_{\text{story}} \cong \frac{L}{h}(V_{\text{link}})_{\text{cl}} = \frac{30}{14} \times 194 = 416 \text{ kips}$$

$$F1 = 416 - 367 = 49 \text{ kips}$$

Figure 7.75 shows the capacity-limited (E_{cl}) loading. By comparison with Figure 7.74, the loading for design of the links, it is clear that the links will be the weakest elements in the load path, as desired. The story shear at each level for the capacity-limited condition is much higher than the shear distribution used to design the links. The links will also be the most ductile elements in the load path, as desired.

The example illustrates the preference to avoid using over-designed links. In particular, the roof link has an excess of reserve capacity, given that it is the same cross-section (W10×49) as the fourth-floor link. This produces higher than desired capacity-limited forces. Any excess link capacity will result in higher capacity-limited force distributions.

This example also illustrates a method for generating the capacity-limited shear distribution required to produce the adjusted shear in the links at each level. The capacity-limited shear distribution is used for the design of the braces, the columns, and the beams outside the central link segment. AISC 341 stipulates that these elements and their connections are to be designed for the capacity-limited load effect. One exception noted in AISC 341 is that the beams outside the link segment may be designed for 0.88 times the capacity-limited load effect.

An effective strategy for EBF design is to:

1. Size the links for strength, limiting any excess shear resistance as much as possible.
2. From the link sizes, determine the capacity-limited loading distribution and size of the braces, columns, and beams outside the links.
3. Using the link-design load distribution (not the capacity-limited load distribution), determine the seismic drift at each level and compare it to the drift capacity based on the link plastic rotation angles.

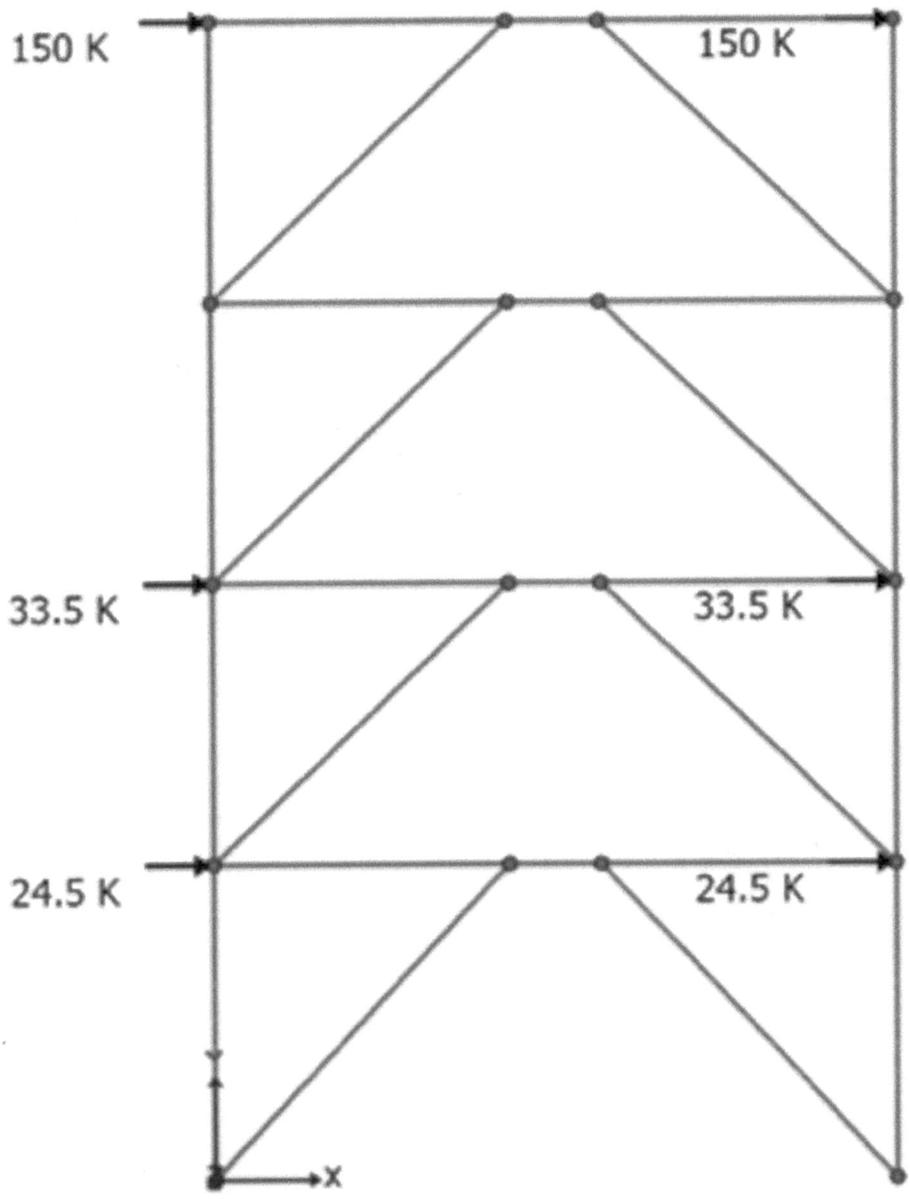

FIGURE 7.75 Capacity-limited loading for example EBF.

4. For any location where displacement capacity is exceeded, stiffen the structure by redesigning the braces and, possibly, the columns. This is preferred to redesigning the links, which would require re-calculation of the capacity-limited force distribution and subsequent re-design of the braces and columns.

Refer to Section F3 of AISC 341 for more detailed discussion and requirements for eccentrically braced frame design.

7.14.8 Lateral Force-Resisting Systems for Bridges

Bridge design for seismic effects requires that a seismic design category, also known as a seismic zone, be established based on the expected ground shaking at the project site.

AASHTO has multiple documents defining seismic loading for bridges. The AASHTO LRFD Bridge Design Specifications (AASHTO, 2020) is force-based in nature and requires the determination of a Seismic Zone. The AASHTO Guide Specifications for LRFD Seismic Bridge Design (AASHTO, 2023) is displacement-based in nature and requires the determination of a Seismic Design Category (SDC). The most recent seismic criteria for bridge design are found in the AASHTO Guidelines for Performance-Based Seismic Design of Highway Bridges (AASHTO, 2023). The performance-based guidelines classify bridges as either critical, ordinary, or recovery and require assessment at two hazard levels. The lower hazard level is ground shaking, having a 100-year mean recurrence interval. The upper-level ground shaking has a 1,000-year mean recurrence interval. In general, bridges designed in accordance with the performance-based guidelines are required to remain fully operational after the lower-level event. Critical bridges are required to remain fully operational even after the upper-level event. Refer to the performance-based guidelines for a complete treatment on the criteria for each bridge classification.

The seismic design category designation depends on the parameter, S_{D1}. The value of S_{D1} is taken as 90% of the maximum product of the period and the corresponding pseudo-spectral acceleration, $T \times PSA$, for periods within a certain range. For sites with an average shear wave velocity greater than 1,450 ft/s, the period range is 1–2 seconds. For sites with an average shear wave velocity less than or equal to 1,450 ft/s, the period range is 1–5 seconds. However, in no case is the value for S_{D1} to be taken less than the pseudo-spectral acceleration at a period of 1 second. Once the appropriate value for S_{D1} has been determined, the seismic design category (SDC) may be determined:

- $S_{D1} < 0.15 \rightarrow$ SDC A (Seismic Zone 1)
- $0.15 \leq S_{D1} < 0.30 \rightarrow$ SDC B (Seismic Zone 2)
- $0.30 \leq S_{D1} < 0.50 \rightarrow$ SDC C (Seismic Zone 3)
- $S_{D1} \geq 0.50 \rightarrow$ SDC D (Seismic Zone 4)

Seismic loading in AASHTO was initially discussed in Section 7.11.

7.14.8.1 Force-based Seismic Design – Ductile Substructure

In force-based seismic design by the AASHTO LRFD Bridge Design Specification (LRFD-BDS) (AASHTO, 2020), a force-reduction factor is first established. This requires that the Owner specify an importance category for each bridge project. These categories include (1) critical, (2) essential, and (3) other. These categories are likely to change soon, but as of May 2023, they are as specified herein. For design by response spectrum analysis, force-reduction factors are summarized in Table 7.37. For design by inelastic response history analysis, $R = 1.0$. With inelastic response history analysis, inelasticity is directly and explicitly modeled.

For connection of ductile columns or pile bent piles to the cap and foundation, it is permitted to use capacity-protection procedures in which overstrength capacity of ductile elements is determined to establish the maximum load that can physically be transmitted to the connection or connected element. The capacity-protection method is preferred and may be used in lieu of the R-factors for connections.

Once determined by an acceptable analysis technique, elastic seismic forces are reduced by the factor, R, and combined for orthogonal effects as follows:

- 100% of the design force in direction 1 combined with 30% of the design force in direction 2
- 100% of the design force in direction 2 combined with 30% of the design force in direction 1

When non-ductile, capacity-protected elements are designed for the expected, overstrength capacity of the ductile elements, such a combination is unnecessary.

For bridges in Seismic Zone 1 with A_S less than 0.05, the design force in any direction is to be taken no less than 15% of the total vertical reaction due to combined (1) permanent loads and (2) live load assumed to act concurrently with the earthquake. For other bridges in Seismic Zone 1, the design force in any direction is not to be taken less than 25% of the total vertical reaction.

For Seismic Zone 2, capacity-protected, non-ductile elements are designed for a force-reduction factor, R, equal to one-half of the value from Table 7.37, but no less than 1.0. Once again, an alternative is to design for expected, overstrength capacity of the ductile elements.

TABLE 7.37
Force-Reduction Factor for Force-based Seismic Design

Substructure	Critical	Essential	Other
Wall-type piers	1.5	1.5	2.0
Concrete pile bents, vertical piles only	1.5	2.0	3.0
Concrete pile bents, batter piles	1.5	1.5	2.0
Single-column bents	1.5	2.0	3.0
Steel and CFST pile bents, vertical piles only	1.5	3.5	5.0
Steel and CFST pile bents, batter piles	1.5	2.0	3.0
Multi-column bents	1.5	3.5	5.0
Connection, superstructure to abutment	0.8	0.8	0.8
Within-span expansion joints	0.8	0.8	0.8
Connection, columns and pile bents to cap	1.0	1.0	1.0
Connection, columns or piles to foundation	1.0	1.0	1.0

For Seismic Zones 3 and 4, hinging, ductile elements (usually the columns in a multi-post bent, or the piles in a pile bent) are first designed using the reduced seismic forces. Capacity-protected, non-ductile elements are then designed for either (1) unreduced elastic seismic forces ($R = 1.0$) or (2) the inelastic hinging forces. Inelastic hinging forces are to be determined using resistance factors, $\phi = 1.3$ for concrete or $\phi = 1.25$ for structural steel, to account for overstrength. This is an estimate of the plastic shear for a bent or pier. The plastic shear is the largest load that may be transmitted from a ductile element to the connected elements.

Section 5.11 of the LRFD-BDS contains specific seismic requirements for concrete elements. For details not covered here, the reader is referred to that section.

Requirements outlined here are generally those applicable to Seismic Zones 3 and 4, as those requirements are more fully developed and appropriate for the seismic design of bridges, in the author's opinion.

For circular columns and piles, Equation 7.85 provides the required volumetric spiral or seismic hoop reinforcement for plastic hinge regions. The core diameter, d_c, is measured to the outside diameter of the spiral or hoop. The pitch, s, is measured vertically to the center of the spiral or hoops. A_{sp} is the cross-sectional area of the spiral or hoop bar. The specified minimum yield strength, f_{yh}, is that for the spiral or hoop bar, not necessarily the same as that for longitudinal bars. The specified concrete strength, f'_c, is used for the calculations, not the expected concrete strength, f'_{ce}. Equation 7.86 from Section 5.6.4.6 of the LRFD-BDS for non-seismic confinement requirements, should always be checked as well. The core area, A_c, is based on the diameter to the outside of the spiral or hoop as well.

Equations 7.87 and 7.88 provide the required confinement reinforcement, A_{sh}, within a spacing equal to s, for rectangular piles and columns. The dimension, h_c, is measured to the outside of the transverse bars in the direction of loading. The spacing, s, is to exceed neither (1) 4 inches, nor (2) one-quarter of the least cross-sectional dimension:

$$\rho_s = \frac{4A_{sp}}{d_c s} \geq 0.12\frac{f'_c}{f_{yh}} \quad (7.85)$$

$$\rho_s = \frac{4A_{sp}}{d_c s} \geq 0.45\left(\frac{A_g}{A_c} - 1\right)\frac{f'_c}{f_{yh}} \quad (7.86)$$

$$A_{sh} \geq 0.12 s h_c \frac{f'_c}{f_{yh}} \quad (7.87)$$

$$A_{sh} \geq 0.30 s h_c \left(\frac{A_g}{A_c} - 1\right)\frac{f'_c}{f_{yh}} \quad (7.88)$$

For either round or rectangular columns and piles, the yield strength of the transverse reinforcement, f_{yh}, is the minimum specified value for the reinforcement used, but is not to exceed 75 ksi.

7.14.8.2 Displacement-based Seismic Design – Ductile Substructure

The most frequently adopted seismic design strategy for typical bridges is to limit the inelastic behavior to plastic hinging in the substructure columns. The superstructure and the foundation system are designed to remain elastic during strong ground shaking using capacity-design procedures. This strategy has historically been referred to as type I seismic design. Figure 7.76 depicts this design strategy.

Displacement-based seismic design involves (1) identification of ductile elements to be designed and detailed for inelastic behavior, (2) ensuring that the displacement

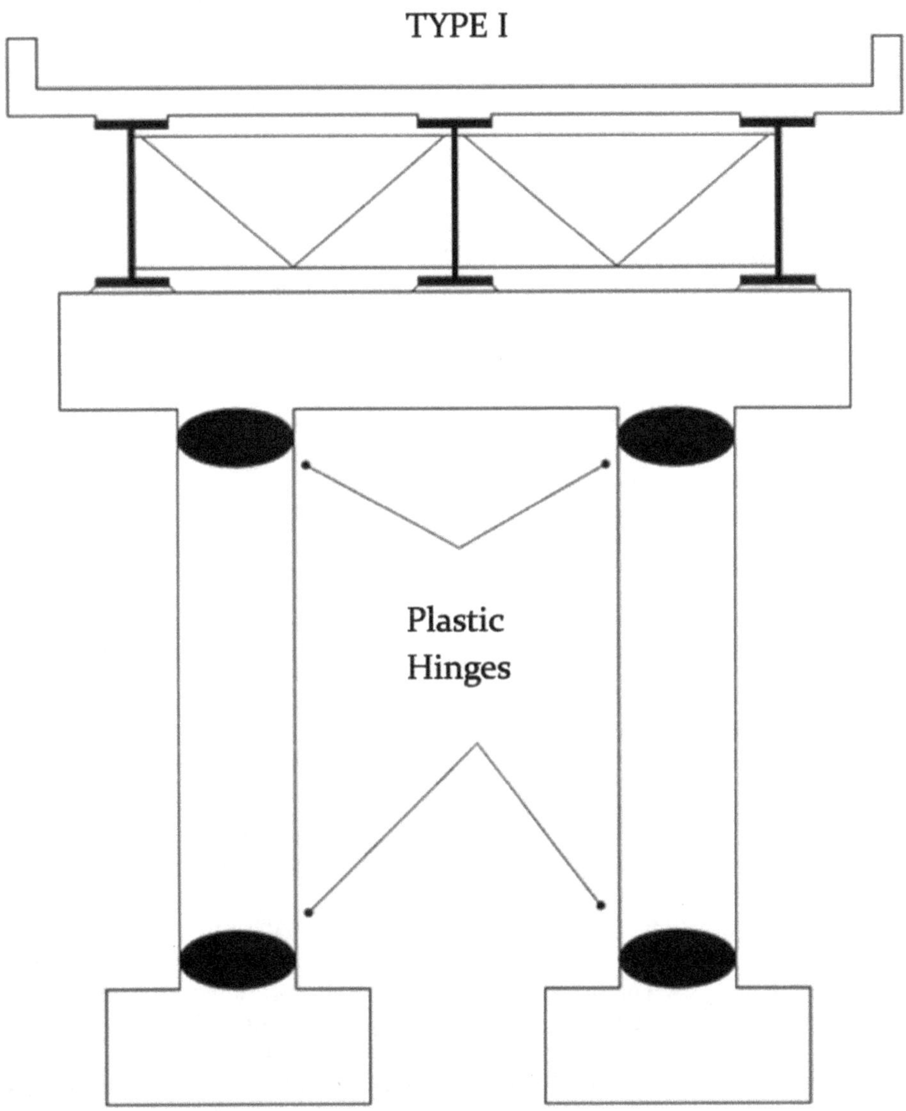

FIGURE 7.76 Ductile substructure (type I) design strategy for bridges.

demand on each ductile element is less than its displacement capacity, and (3) capacity protection of elements intended to remain essentially elastic during strong ground shaking.

Displacement ductility is defined as the maximum displacement experienced by an element divided by the yield displacement of the element. For pile bent design by the displacement-based provisions of the AASHTO Guide Specification for LRFD Seismic Bridge Design (LRFD-GS) (AASHTO, 2011), the displacement ductility demand, μ_D, is not to exceed 4. For single-column bents, μ_D is limited to 5, and for multi-column bents with above-ground hinging, μ_D is limited to 6.

Displacement demand may be determined by (1) equivalent static analysis with displacement amplification for short-period structures, (2) multi-mode elastic response spectrum analysis with displacement amplification for short-period structures, or (3) inelastic response history analysis. Orthogonal component combination for methods (1) and (2) are based on the 100-30-30 rule, while orthogonal component interaction is inherently accounted for in inelastic response history analysis, when three-dimensional modeling is adopted. Refer to the LRFD-GS, Section 4.2, for guidance on permissible analysis methods for various bridge complexity levels.

For equivalent static and elastic response spectrum methods, short-period displacement amplification is required for natural modes having periods less than or equal to $1.25T_S$. See Section 7.10.1 on earthquake loading in ASCE 7 for definition of T_S. The required amplification, R_d, of displacement demand is given by Equation 7.89. Design response spectra are typically based on damping equal to 5% of critical. In rare cases for which larger damping is appropriate, Equation 7.90 is provided in the LRFD-GS for modification of the design response spectrum across all periods by the factor, R_D:

$$R_d = \left(1 - \frac{1}{\mu_D}\right)\left(\frac{T}{1.25T_S}\right) + \frac{1}{\mu_D} \tag{7.89}$$

$$R_D = \left(\frac{0.05}{\xi}\right)^{0.40} \tag{7.90}$$

For Equation 7.89, μ_D is equal to 2 for Seismic Design Category B, 3 for Seismic Design Category C, and determined by analysis for Seismic Design Category D. For Seismic Design Category D, μ_D may conservatively be taken equal to 6 in Equation 7.89.

Second-order $P\Delta$ effects may be ignored in structural analysis for earthquake loading whenever Equation 7.91 is satisfied for concrete substructures, or Equation 7.92 for steel substructures. The unfactored dead load is P_{dl}. M_p is the idealized plastic moment determined using expected, rather than specified minimum, material properties. M_n is the nominal moment capacity based on nominal properties. Presumably, Δ_r is the first-order displacement between points of contraflexure and plastic hinging (approximated by one-half of the total displacement for a rigid frame with hinges at top and bottom of columns):

$$P_{\text{dl}}\Delta_r \leq 0.25M_p \tag{7.91}$$

$$P_{\text{dl}}\Delta_r \leq 0.25M_n \tag{7.92}$$

Equation 7.93 for analytical plastic hinge length, L_P, is appropriate for reinforced concrete columns framing into oversized shafts, footings, cased shafts, and bent caps. In the equation, L is the distance from the point of maximum moment to the point of contraflexure, not necessarily equal to the clear column height. The diameter, d_{bl}, is that for the longitudinal bars in the column. For reinforced and prestressed concrete piling, and for cast-in-drilled-hole shafts, Equation 7.94 provides the analytical plastic hinge length. H' is the distance from ground surface to point of above-ground contraflexure. D^* is the diameter for a circular member, or the cross-section dimension in the direction of loading otherwise. For concrete-filled steel tube pipe piles of diameter D, Equation 7.95 is specified in the LRFD-GS:

$$L_P = 0.08L + 0.15f_{\text{ye}}d_{\text{bl}} \geq 0.30f_{\text{ye}}d_{\text{bl}} \tag{7.93}$$

$$L_P = 0.1H' + D^* \leq 1.5D^* \tag{7.94}$$

$$L_P = 0.1H' + 1.25D \leq 2.0D \tag{7.95}$$

Plastic hinge lengths are used in conjunction with yield and ultimate curvature estimates to establish displacement capacities in a hand-calculation-based pushover analysis. Pushover analysis may also be accomplished with computer software. For hand-calculation-based pushover analysis, displacement capacity may be estimated using Equations 7.96 through 7.115. Definitions for the parameters appearing in the equations are given after the equations:

$$f_l' = \frac{1}{2}C_e\rho_v f_{\text{yh}} \rightarrow \text{lateral confining stress on concrete} \tag{7.96}$$

$$k_e = \frac{\left(1 - \dfrac{s'}{2d_s}\right)^2}{1 - \rho_{\text{cc}}} \rightarrow \text{circular hoop effectiveness coefficient} \tag{7.97}$$

$$k_e = \frac{1 - \dfrac{s'}{2d_s}}{1 - \rho_{cc}} \rightarrow \text{circular spiral effectiveness coefficient} \tag{7.98}$$

$$k_e = \frac{\left(1 - \sum \dfrac{(w_i')^2}{6b_c d_c}\right)\left(1 - \dfrac{s'}{2b_c}\right)\left(1 - \dfrac{s'}{2d_c}\right)}{1 - \rho_{\text{cc}}} \rightarrow \text{rectangular hoops} \tag{7.99}$$

$$\rho_{\text{cc}} = \frac{A_s}{A_c} = \frac{\text{Area of longitudinal reinforcement}}{\text{Area of core to centerline of hoop or spiral}} \tag{7.100}$$

$$\rho_v = \frac{A_{sp}\pi d_s}{\frac{\pi}{4}d_s^2 s} = \frac{4A_{sp}}{d_s s} \rightarrow \text{circular hoops or spirals} \tag{7.101}$$

$$\rho_v = \frac{A_{sx}}{sd_c} \rightarrow \text{rectangular hoops} \tag{7.102}$$

$$f'_{cc} = f'_{co}\left(-1.254 + 2.254\sqrt{1 + \frac{7.94 f'_l}{f'_{co}}} - 2\frac{f'_l}{f'_{co}}\right) \tag{7.103}$$

$$\varepsilon_{cc} = 0.002\left[1 + 5\left(\frac{f'_{cc}}{f'_{co}} - 1\right)\right] \tag{7.104}$$

$$\phi_y \cong 2.25\,{}^{\varepsilon_y}\!/_{D} \text{ for circular concrete columns} \tag{7.105}$$

$$\phi_y \cong 2.10\,{}^{\varepsilon_y}\!/_{D} \text{ for rectangular concrete columns} \tag{7.106}$$

$$\frac{c}{D} \cong 0.20 + 0.65\frac{P}{f'_{ce}A_g} \tag{7.107}$$

$$\varepsilon_{cu} = 0.004 + \frac{1.4\rho_v f_{yh}\varepsilon_{su}}{f'_{cc}} \tag{7.108}$$

$$\varepsilon_{su} = 0.06 \tag{7.109}$$

$$\Delta_y = \frac{1}{3}\phi_y\left(L_c + L_{SP}\right)^2 \tag{7.110}$$

$$\Delta_P = \left(\phi_u - \phi_y\right)L_P\left(L_c + L_{SP} - \frac{L_P}{2}\right) \tag{7.111}$$

$$\Delta_u = \Delta_y + \Delta_P \tag{7.112}$$

$$L_{SP} = 0.15 f_{ye} d_{bl} \tag{7.113}$$

$$L_P = kL_c + L_{SP} \geq 2L_{SP} \tag{7.114}$$

$$k = 0.20\left(\frac{f_u}{f_y} - 1\right) \leq 0.08 \tag{7.115}$$

- b_c is the width of a rectangular section, measured to the center of confining bars
- d_c is the depth of a rectangular section, measured to the center of confining bars
- D is the diameter of a round column or the depth of a rectangular column
- f'_{co} is the specified concrete strength for unconfined concrete
- f'_{cc} is the compressive strength of confined concrete
- f_l is the lateral confining stress

- f_y is the yield strength of the longitudinal bars
- f_u is the tensile strength of the longitudinal bars
- f_{yh} is the yield strength of the confining reinforcement
- w' is the clear distance between adjacent longitudinal bars in a rectangular section
- d_s is the core diameter of a circular section, measured to the centerline of the hoop or spiral
- s is the center-to-center spacing of hoops or spirals
- s' is the clear distance between adjacent hoops or spirals
- A_c is the core area, measured to the center of hoops or spirals in a round concrete section
- A_{sx} is the total area of transverse reinforcing bars crossing a plane perpendicular to the direction of displacement
- L_c is the distance from the point of contraflexure to the point of maximum moment
- L_p is the plastic hinge length
- L_{sp} is the depth of strain penetration
- ϕ_y is the yield curvature
- ϕ_u is the ultimate curvature
- Δ_y is the yield displacement
- Δ_p is the plastic displacement capacity
- Δ_u is the ultimate displacement capacity
- c is the distance from the compression face to the neutral axis
- e_{cu} is the maximum usable concrete strain for confined concrete
- e_{su} is the maximum usable steel strain

The plastic hinge region for reinforced concrete columns and piles, within which enhanced lateral confinement reinforcement is to be provided, is taken as the greater of:

- 1.5 times the dimension in the direction of loading
- The region where moment demand exceeds 75% of the plastic moment
- The analytical plastic hinge length, L_P

Minimum support lengths at expansion joints are determined from Equation 7.116 for Seismic Design Category A. L is the length, in feet, of bridge to the adjacent expansion joint or the end of the bridge. N is the minimum required support length, in inches. S is the skew of the deck to substructure alignment, in degrees. H is the average height, in ft, of columns supporting the bridge deck at piers from abutment to abutment:

$$N = (8 + 0.02L + 0.08H)\left(1 + 0.000125S^2\right) \qquad (7.116)$$

For Seismic Design Categories B and C, the length given by Equation 7.116 is to be increased by a factor of 1.5.

For Seismic Design Category D, the required support length is a function of the displacement demand from the structural analysis (Δ_{eq}), and is given by Equation 7.117, but not to be taken less than 1.5 times the value from Equation 7.116:

$$N = \left(4 + 1.65\Delta_{eq}\right)\left(1 + 0.000125S^2\right) \geq 24 \tag{7.117}$$

For structural analysis, abutment longitudinal stiffness (K_{eff}) and strength (P_p) may be neglected, with intermediate piers designed to resist all seismic effects, or may be included (when permitted by the Owner) as given in Equations 7.118 and 7.119:

$$P_p = H_w W_w p_p \tag{7.118}$$

$$K_{eff1} = \frac{P_p}{F_w H_w} \tag{7.119}$$

- K_{eff} = abutment stiffness, kips/ft
- P_p = passive capacity, kips
- p_p = presumptive passive pressure, ksf
- H_w = backwall height, ft
- W_w = backwall width, ft
- F_w = factor ranging from 0.01 (dense sand) to 0.05 (compacted clay)

Unless special attention, beyond that normally provided for backfill, is introduced into the plans and specifications, 70% of the presumptive passive pressure, p_p, should be used in the analysis. Refer to the LRFD-GS, Section 5.2.3.1, for additional information. Presumptive pressure, if used in lieu of detailed analysis procedures, requires that the backfill be compacted to a dry density greater than 95% of the maximum and is specified in the LRFD-GS as follows:

- For cohesionless, non-plastic backfill, p_p may be taken equal to $2H_w/3$ ksf/ft of wall length.
- Provided that the estimated undrained shear strength is greater than 4 ksf, p_p for cohesive backfill may be taken equal to 5 ksf.

Equation 7.119 provides an initial stiffness estimate, which may be used in a response spectrum or response history analysis in cases for which abutment stiffness and strength are to be used to resist seismic forces. From the initial structural analysis, the longitudinal forces at the abutments will be available. Should this force exceed P_p, the initial spring stiffness should be softened and the analysis re-run. The iterative procedure should be performed until the assumed stiffness is consistent with the computed stiffness.

Displacement-based and force-based designs both require attention to capacity-design principles. Ductile elements are detailed to ensure inelastic deformation capacities more than the estimated required deformation during strong ground shaking. Other elements are designed for a higher force level, thus ensuring their essentially elastic behavior during the event. For bridge substructures, when column plastic hinging is the design strategy, overstrength plastic shear calculations are made, and all other elements are designed for the expected, overstrength plastic shear.

Some of these capacity-design provisions from the AASHTO LRFD-GS are summarized here. Readers are referred to the LRFD-GS for a complete treatment of these principles and requirements.

Plastic hinging forces are to be computed using expected (not minimum) material strengths with an additional overstrength factor. The overstrength factor, λ_{mo}, is determined as follows:

- $\lambda_{mo} = 1.4$, reinforced concrete hinging columns with A 615 reinforcing
- $\lambda_{mo} = 1.2$, reinforced concrete with A 706 reinforcing
- $\lambda_{mo} = 1.2$, structural steel hinging columns

Expected material strengths are $f'_{ce} = 1.3 f'_c$ for concrete and 68 ksi for the yield stress, f_{ye}, of steel reinforcing bars, whether A615 Grade 60 or A706 Grade 60. Expected tensile strength, f_{ue}, is 95 ksi for both bar specifications. For hinging columns in Seismic Design Category D, A706 reinforcement is required over A615.

Strain limits in reinforcing are summarized below, with recommended ultimate tensile strains on the order of two-thirds to three-fourths of actual minimum values for added safety. The onset of strain hardening is defined by ε_{sh}, and the ultimate tensile strain by ε_{su}^R.

For the onset of strain hardening:

- $\varepsilon_{sh} = 0.0150$, A 615 and A 706 bars, #3-#8
- $\varepsilon_{sh} = 0.0125$, A 615 and A 706 bars, #9
- $\varepsilon_{sh} = 0.0115$, A 615 and A 706 bars, #10-#11
- $\varepsilon_{sh} = 0.0075$, A 615 and A 706 bars, #14
- $\varepsilon_{sh} = 0.0050$, A 615 and A 706 bars, #18

For bar fracture:

- $\varepsilon_{su}^R = 0.090$, A 706 bars, #4–10
- $\varepsilon_{su}^R = 0.060$, A 615 bars, #4–10
- $\varepsilon_{su}^R = 0.060$, A 706 bars, #11–18
- $\varepsilon_{su}^R = 0.040$, A 615 bars, #11–18

Limiting compressive strains in the concrete core are typically determined using the Mander model for confined concrete (see Equations 7.96 and following) and depend on the amount of transverse reinforcement provided in the form of hoops, spirals, or rectangular ties. Core compressive strains at in-ground plastic hinges, when permitted, should be limited to 0.02.

Determination of the plastic shear for a multi-column bent or pier is an iterative procedure. Section analysis software is a requirement for such analyses.

Capacity-design provisions for footings may be found in the AASHTO LRFD-GS, Sections 6.4.5, 6.4.6, and 6.4.7. Capacity-design provisions for non-integral bent caps are extensive and are found in Sections 8.12 and 8.13 of the LRFD-GS. The reader is referred to these provisions for detailed design requirements.

7.14.8.3 Performance-Based Seismic Design of Bridges

The ***AASHTO Guidelines for Performance-Based Seismic Bridge Design*** (AASHTO, 2023) recognize the need to provide for differing levels of bridge performance during increasingly severe seismic events. The guidelines represent the first national implementation of performance-based seismic design for bridges in the United States with the goal of facilitating the selection of appropriate levels of performance for a given bridge. Three performance levels are included in the guidelines: life safety, operational, and fully operational. Three operational categories are also included: critical, recovery, and ordinary.

Successful performance-based seismic design requires an established relationship between those response quantities (called engineering design parameters) typically determined from structural analysis and the desired performance level.

Table 7.38 summarizes the relationship between engineering design parameters and the anticipated performance level of a bridge for reinforced concrete columns and concrete-filled steel tubes.

- ε_s = strain in steel bar reinforcing
- ε_c = compressive strain in concrete
- D = tube diameter
- t = tube thickness
- See Section 7.14.8.2 for the definition of the other parameters in the table.

Unlike other bridge design specifications from AASHTO, performance-based seismic design provisions require assessment at two hazard levels. Bridges are required to remain fully operational at the lower-level ground shaking. Performance at the upper-level ground shaking depends on the bridge classification (critical, recovery, or ordinary).

TABLE 7.38
Performance-Based Design: Engineering Design Parameters

Engineering Design Parameters (EDP's)	Performance Level		
	PL1 Life Safety	**PL2 Operational**	**PL3 Fully Operational**
Longitudinal reinforcement tensile strain (RC)	$\varepsilon_s = 0.032 + 790\rho_s \frac{f_{\text{yhe}}}{E_s} - 0.14\frac{P}{f'_{\text{ce}}A_g}$	$\varepsilon_s = 0.8(\varepsilon_s)_{\text{PL1}}$	0.010
Concrete compressive strain (RC)	$\varepsilon_c = 1.4(\varepsilon_c)_{\text{PL2}}$	$\varepsilon_c = 0.004 + 1.4\frac{\rho_v f_{\text{yh}}\varepsilon_{\text{su}}}{f'_{\text{cc}}}$	0.004
Steel tube tensile strain (RCFST)	0.025	$\varepsilon_s = 0.021 - \frac{D/t}{9{,}100} \geq \varepsilon_y$	ε_y
Concrete compressive strain (RCFST)	NA	NA	NA

7.14.8.4 Seismic Design Strategy – Ductile Superstructure

A less-used, but still available, method of providing seismic resistance is to limit the inelastic response to ductile cross-frames at the bridge supports. This is called Type II design and is depicted in Figure 7.77. In the Figure, an eccentrically braced cross-frame is shown. Such conFigurations may prove highly efficient for transverse movement of the bridge superstructure. One of the difficulties with such systems is accommodating the response in the longitudinal direction (into the page in Figure 7.77).

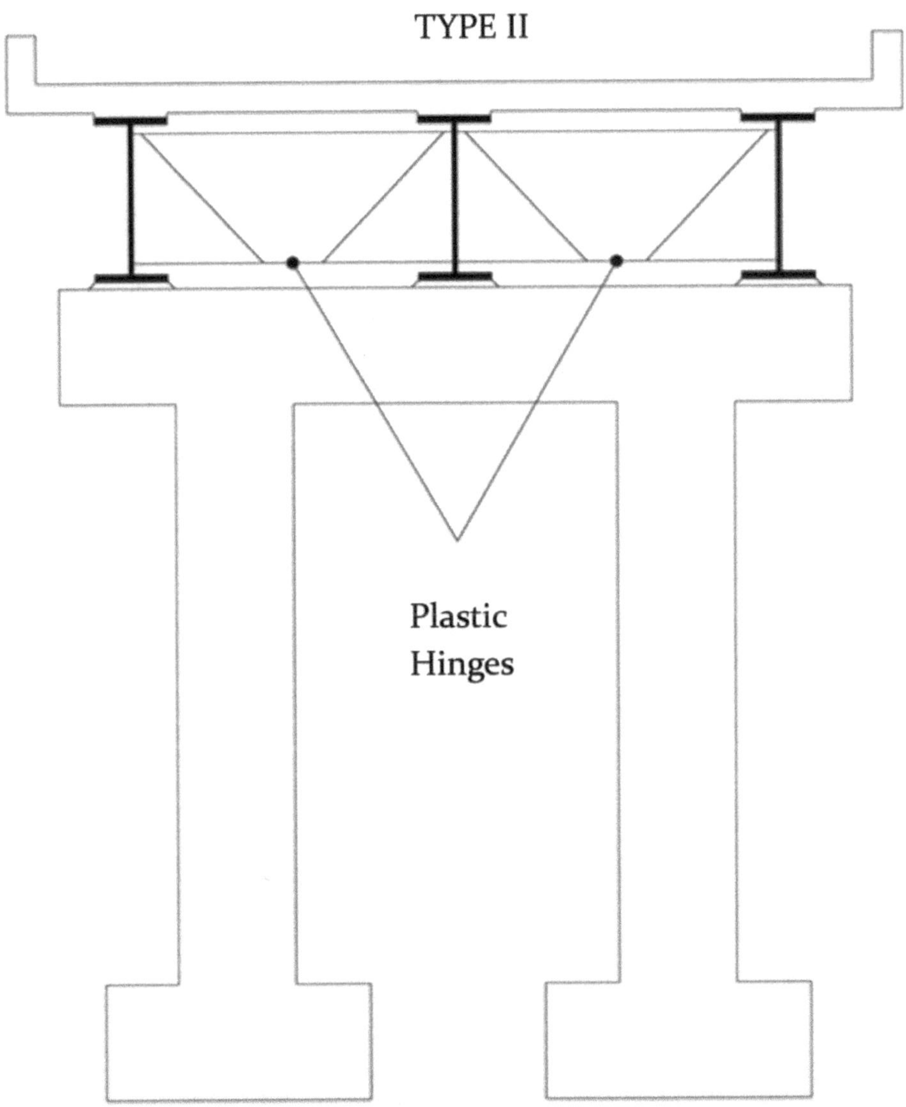

FIGURE 7.77 Ductile cross-frame (type II) seismic design strategy for bridges.

7.14.8.5 Seismic Isolation of Bridges

Another option for seismic resistance is to provide isolation bearings for the bridge girders. This is referred to as Type III design. All inelastic behavior occurs in the bearings. The superstructure above the bearings and the substructure and foundation system below the bearings remain elastic through capacity-design principles. Figure 7.78, with the bearings in a deflected position, illustrates the concept of seismic isolation as applied to bridges.

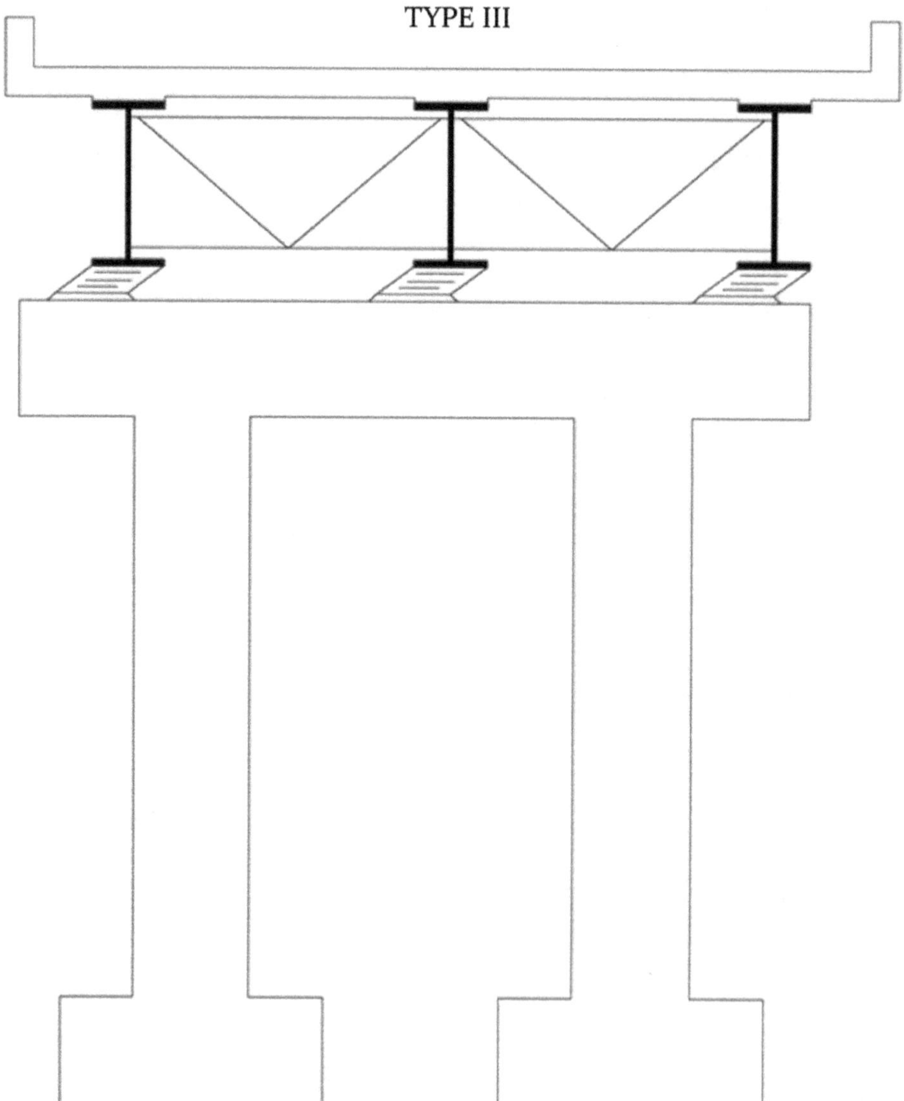

FIGURE 7.78 Seismic isolation (type III) of bridges.

Commonly used seismic isolation devices for bridges include lead-rubber bearings (LRB) and friction-pendulum systems (FPS). LRB isolation devices were used on the Interstate 40 bridge over State Route 5 in Madison County, Tennessee (see Figure 7.79). FPS isolation devices were used in the seismic retrofit of the main 900-ft arch spans for Interstate 40 over the Mississippi River (see Figures 7.80 and 7.81).

The LRB bearings shown in Figure 7.79 consist of intermittent layers of elastomer and steel with a central lead plug, 3.75 inches in diameter. The bearings are 8.75 inches high and 22.5 inches in diameter and accommodate an anticipated seismic lateral displacement of 8.3 inches.

FIGURE 7.79 Lead-rubber-bearing isolation devices for I-40 over SR-5.

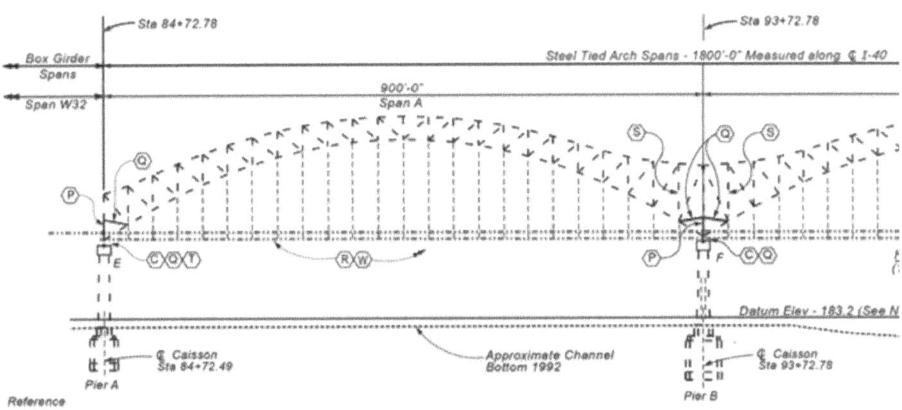

FIGURE 7.80 Main arch spans for I-40 over the Mississippi River.

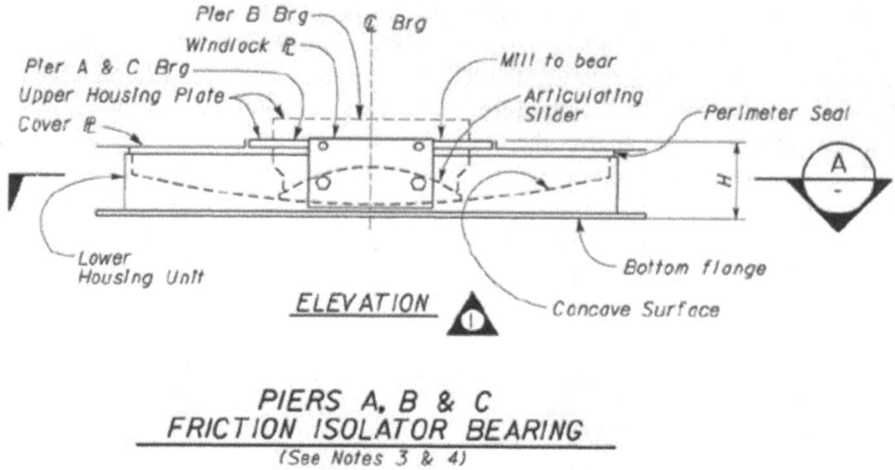

FIGURE 7.81 FPS isolation device for I-40 over the Mississippi River.

The FPS isolation devices shown in Figure 7.81 include a 4′–10″ radius lower housing unit with a 1′–10″ bearing height, *H*. The bearings accommodate 18.75 inches of anticipated lateral seismic displacement at the center pier and 27.25 inches of lateral displacement and the end piers.

Bibliography

Aas, A. & Sinha, S. K., 2023. Seismic Site Characterization Using MASW and Correlation Study between Shear Wave Velocity and SPT–N. *Journal of Applied Geophysics*, 215, pp. 1–11.

AASHTO, 2011. *Guide Specifications for LRFD Seismic Bridge Design (*2nd edition), Washington, DC: American Association of State Highway and Transportation Officials.

AASHTO, 2014. *Guide Specifications for Seismic Isolation Design (*4th edition), Washington, DC: American Association of State Highway and Transportation Officials.

AASHTO, 2017. *AASHTO LRFD Bridge Design Specifications (*8th edition), Washington, DC: American Association of State Highway and Transportation Officials.

AASHTO, 2020. *LRFD Bridge Design Specifications (*9th edition), Washington, DC: American Association of State Highway and Transportation Officials.

AASHTO, 2023a. *Guide Specifications for LRFD Seismic Bridge Design (*3rd Edition), Washington, DC: American Association of State Highway and Transportation Officials.

AASHTO, 2023b. *Guidelines for Performance-Based Seismic Design of Highway Bridges*, Washington, DC: American Association of State Highway and Transportation Officials.

AISC, 2016. *ANSI/AISC 341: Seismic Provisions for Structural Steel Buildings*, Chicago, IL: American Institute of Steel Construction.

AISC, 2016. *ANSI/AISC 358: Prequalified Connections for Special and Intermediate Steel Moment Frames for Seismic Applications*, Chicago, IL: American Institute of Steel Construction.

AISC, 2016. *ANSI/AISC 360: Specification for Structural Steel Buildings*, Chicago, IL: American Institute of Steel Construction.

AISC, 2018. *Specification for Safety-Related Steel Structures for Nuclear Facilities*, Chicago, IL: American Institute of Steel Construction.

Alhuay-Leon, C. G. & Trejo-Norena, P. C., 2021. The Empirical Correlation between Shear Wave Velocity and Penetration Resistance for the Eolian Sand Deposits in the City of Olmos Perui. *DYNA*, 88(217), pp. 247–255.

American Society of Civil Engineers, 2011. *Standard 59-11: Blast Protection of Buildings*, Reston, VA: ASCE.

American Society of Civil Engineers, 2016. *ASCE* 4-16*: Seismic Analysis of Safety-Related Nuclear Structures*, Reston, VA: ASCE.

American Society of Civil Engineers, 2017. *ASCE* 41-17*: Seismic Evaluation and Retrofit of Existing Buildings*, Reston, VA: ASCE.

American Society of Civil Engineers, 2017. *ASCE* 7-16*: Minimum Design Loads and Associated Criteria for Buildings and Other Structures*, Reston, VA: ASCE.

American Society of Civil Engineers, 2019. *ASCE* 43-19*: Seismic Design Criteria for Structures, Systems, and Components in Nuclear Facilities*, Reston, VA: ASCE.

American Society of Civil Engineers, 2022. *ASCE* 7-22*: Minimum Design Loads and Associated Criteria for Buildings and Other Structures*, Reston, VA: ASCE.

American Society of Civil Engineers, 2024. *ASCE Hazard Tool.* [Online] Available at: https://ascehazardtool.org/. [Accessed 20 February 2024].

Assadollahi, C., Pezeshk, S. & Campbell, K., 2023. A Seismological Method for Estimating the Long-period Transition Period TL in the Seismic Building Code. *Earthquake Spectra*, 39, pp. 1–21.

Atkinson, G. & Beresnev, I. A., 2002. Ground Motions at Memphis and St. Louis from M 7.5-8.0 Earthquakes in the New Madrid Seismic Zone. *Bulletin of the Seismological Society of America*, 92(3), pp. 1015–1024.

Atkinson, G. M., 2009. Earthquake Time Histories Compatible with the 2005 NBCC Uniform Hazard Spectrum. *Canadian Journal of Civil Engineering*, 36(6), pp. 991–1000.

Barrette, P. et al., 2017. *Assessing Ice Action on Bridges in the Context of Climate Change: Prospective Approach*, Ottawa, ON: National Research Council Canada.

Bommer, J. J., Elnashai, A. S. & Weir, A. G., 2000. Compatible Acceleration and Displacement Spectra for Seismic Design Codes. *12th World Conference on Earthquake Engineering*, Auckland, New Zealand, Sunday 30 January to Friday 4 February 2000, pp. 1–8.

Boore, D. M., 2001. Effect of Baseline Corrections on Displacements and Response Spectra for Several Recordings of the 1999 Chi-Chi, Taiwan, Earthquake. *Bulletin of the Seismological Society of America*, 91(5), pp. 1199–1211.

Boore, D. M., 2020. *TSPP – A Collection of FORTRAN Programs for Processing and Manipulating Time Series*, Reston, VA: U.S. Geological Survey Open-File Report 2008-1111.

Boore, D. M. & Akkar, S., 2003. Effect of Causal and Acausal Filters on Elastic and Inelastic Response Spectra. *Earthquake Engineering and Structural Dynamics*, 32, pp. 1729–1748.

Boore, D. M. & Bommer, J. J., 2005. Processing of Strong-Motion Accelerograms: Needs, Options and Consequences. *Soil Dynamics and Earthquake Engineering*, 25, pp. 93–115.

Boore, D. M. & Kishida, T., 2017. Relations between Some Horizontal-Component Ground-Motion Intensity Measures Used in Practice. *Bulletin of the Seismological Society of America*, 107(1), pp. 334–343.

Booth, E., 2007. The Estimation of Peak Ground-motion Parameters from Spectral Ordinates. *Journal of Earthquake Engineering*, 11(1), pp. 13–32.

Bozorgnia, Y., 2023. *NGA-Subduction Project.* [Online] Available at: https://www.risksciences.ucla.edu/nhr3/nga-subduction. [Accessed March 2023].

Brown, N. K., Kowalsky, M. J. & Nau, J. M., 2015. Impact of D/t on Seismic Behavior of Reinforced Concrete Filled Steel Tubes. *Journal of Constructional Steel Research*, 107, pp. 111–123.

Bulson, P. S., 1997. *Explosive Loading of Engineering Structures (*1st edition*)*, London: E & FN Spon.

Charney, F. A., Heausler, T. F. & Marshall, J. D., 2020. *Guide to the Seismic Load Provisions of ASCE* 7-16 *(*1st e*dition)*, Reston, VA: ASCE.

Chopra, A. K., 2016. *Dynamics of Structures: Theory and Application to Earthquake Engineering (*5th edition), Harlow: Pearson.

Civil Engineering Faculty – The University of Chile, 2020. *Earthquakes of Chile.* [Online] Available at: https://terremotos.ing.uchile.cl/. [Accessed 24 April 2023].

Clough, R. W. & Penzien, J., 1975. *Dynamics of Structures (*1st edition), New York: McGraw-Hill.

Collins, P. S., 2013. *Ice Loads on the Confederation Bridge Piers*, St. Johns, NL, Canada: Journal of Undergraduate Engineering Research and Scholarship.

Consortium of Organizations for Strong-Motion Observation Systems (COSMOS), 2012. *Strong Motion Virtual Data Center.* [Online] Available at: https://www.strongmotion-center.org/vdc/scripts/default.plx. [Accessed 24 April 2023].

Corchette, V., 2010. The Analysis of Accelerograms for the Earthquake Resistant Design of Structures. *International Journal of Geosciences*, 1, pp. 32–37.

Defense, U. S. D. o. D., 2010. *UFC 4-023-03: Design of Buildings to Resist Progressive Collapse*, Washington, DC: US Army Corps of Engineers, Naval Facilities Engineering Command, Air Force Civil Engineer Support Agency.

Dikmen, U., 2009. Statistical Correlations of Shear Wave Velocity and Penetration Resistance for Soils. *Journal of Geophysics and Engineering*, 6, pp. 61–72.

European Infrastructure for Seismic Waveform Data, 2016. *ORFEUS.* [Online] Available at: https://www.orfeus-eu.org/data/strong/. [Accessed 24 April 2023].

Fernandez, J. A., 2007. *Numerical Simulation of Earthquake Ground Motions in the Upper Mississippi Embayment*, Atlanta: Georgia Institute of Technology.

Fernandez, J. A. & Rix, G. J., 2006. Soil Attenuation Relationships and Seismic Hazard Analyses in the Upper Mississippi Embayment. *8th U.S. National Conference on Earthquake Engineering*, San Francisco, CA, 18–22 April 2006.

Field, E. H., Jordan, T. H. & Cornell, C. A., 2003. OpenSHA: A Developing Community – Modeling Environment for Seismic Hazard Analysis. *Seismological Research Letters*, 74(4), pp. 406–419.

Fulmer, S. J., Kowalsky, M. J. & Nau, J. M., 2012. Reversed Cyclic Flexural Behavior of Spiral DSAW and Single Seam ERW Steel Pipe Piles. *ASCE Journal of Structural Engineering*, 138(9), pp. 1099–1109.

Gilsanz, R. et al., 2013. *Design of Blast Resistant Structures*, Chicago, IL: American Institute of Steel Construction.

Graizer, V. & Kalkan, E., 2015. *Update of the Graizer-Kalkan Ground-Motion Prediction Equations for Shallow Crustal Continental Earthquakes*, Reston, VA: USGS Open-File Report 2015-1009.

Hashash, Y. M. A., Tsai, C.-C., Phillips, C. & Park, D., 2008. Soil-Column Depth-Dependent Seismic Site Coefficients and Hazard Maps for the Upper Mississippi Embayment. *Bulletin of the Seismological Society of America*, 98(4), pp. 2004–2021.

Huff, T., 2018. Inelastic Seismic Displacement Amplification for Bridges: Dependence upon Various Intensity Measures. *ASCE Practice Periodical on Structural Design and Construction*, 23(1). https://doi.org/10.1061/(ASCE)SC.1943-5576.0000355

Huff, T. & Pezeshk, S., 2016. Inelastic Displacement Spectra for Bridges Using the Substitute-Structure Method. *ASCE Practice Periodical on Structural Design and Construction*, 21(2). https://doi.org/10.1061/(ASCE)SC.1943-5576.0000279

Kennedy, R. P. et al., 1990. *UCRL 15910: Design and Evaluation Guidelines for Department of Energy Facilities Subjected to Natural Phenomena Hazards*, Livermore, CA: Lawrence Livermore National Laboratory.

Kottke, A. R. & Rathje, E. M., 2012. *Technical Manual for SigmaSpectra.* [Online] Available at: https://github.com/arkottke/sigmaspectra/blob/master/manual/manual.pdf. [Accessed 28 April 2023].

Lehman, D. E. & Roeder, C. W., 2012. *An Initial Study into the Use of Concrete Filled Steel Tubes for Bridge Piers and Foundation Connections*, Seattle: University of Washington.

Lehman, D. E. & Roeder, C. W., 2012. *Rapid Construction of Bridge Piers with Improved Seismic Performance*, Seattle: University of Washington.

Liu, C. & Wang, T.-L., 2001. Statewide Vessel Collision Design for Bridges. *ASCE Journal of Bridge Engineering*, 6(3), pp. 213–219.

Malhotra, P. K., 2006. Smooth Spectra of Horizontal and Vertical Ground Motions. *Bulletin of the Seismological Society of America*, 96(2), pp. 506–518.

Marsh, M. L. & Stringer, S. J., 2013. *NCHRP Synthesis 440: Performance-Based Seismic Bridge Design*, Washington, DC: Transportation Research Board.

Mazzoni, S., 2022. *NGA-Subduction Portal: Ground-Motion Record Selection and Download.* [Online] Available at: https://www.risksciences.ucla.edu/nhr3/nga-subduction/gmportal. [Accessed 24 April 2023].

Montejo, L., 2021. Response Spectral Matching of Horizontal Ground Motion Components to an Orientation-Independent Spectrum (RotDnn). *Earthquake Spectra*, 37(2), pp. 1127–1144.

Montejo, L. A., Gonzalez-Roman, L. A. & Kowalsky, M. J., 2012. Seismic Performance of Reinforced Concrete-Filled Steel Tube Pile/Column Bridge Bents. *Journal of Earthquake Engineering*, 16(3), pp. 401–424.

Moon, S., Hashash, Y. M. A. & Park, D., 2016. USGS Hazard Map Compatible Depth-Dependent Seismic Site Coefficients for the Upper Mississippi Embayment. *KSCE Journal of Civil Engineering*, 21(1), pp. 1–12.

Murphy, T. P. et al., 2020. *NCHRP Research Report 949: Proposed AASHTO Guidelines for Performance-Based Seismic Bridge Design*, Washington, DC: The National Academies of Sciences, Engineering, and Medicine.

NEHRP Consultants Joint Venture, 2011. *NIST GCR 11-917-15: Selecting and Scaling Earthquake Ground Motions for Performing Response-History Analyses*, Washington, DC: US Department of Commerce.

NEHRP Consultants Joint Venture, 2011. *Selecting and Scaling Earthquake Ground Motions for Performing Response-History Analyses*, Gaithersburg, MD: National Institute of Standards and Technology.

New Zealand Government, 2017. *New Zealand Strong-Motion Database.* [Online] Available at: https://www.geonet.org.nz/data/supplementary/nzsmdb. [Accessed 24 April 2023].

Pacific Earthquake Engineering Research Center, 2013. *PEER Ground Motion Database.* [Online] Available at: https://ngawest2.berkeley.edu/. [Accessed 24 April 2023].

Perez-Santisteban, I., Martin, A. M., Gorosab, A. C. & Fonticiella, J. M. R., 2016. Empirical Correlation of Shear Wave Velocity (Vs) with SPT of Soils in Madrid. *First Break*, 34(8), pp. 87–92.

Priestley, M. J. N., Calvi, G. M. & Kowalsky, M. J., 2007. *Displacement Based Seismic Design of Structures*, Pavia, Italy: IUSS Press.

Priestley, M. J. N. & Grant, D. N., 2005. Viscous Damping in Seismic Design and Analysis. *Journal of Earthquake Engineering*, 9(SPEC02), pp. 229–255.

Priestley, M. J. N., Seible, F. & Calvi, G. M., 1996. *Seismic Design and Retrofit of Bridges*, New York: Wiley-Interscience.

Ramon, L., Huff, T. & VandenBerge, D., 2024. Long-Period Transition for Subduction Earthquake Spectra. *ASCE Practice Periodical on Structural Design and Construction*, 29(2). https://doi.org/10.1061/PPSCFX.SCENG-1400

Rodtang, E., Alfredsen, K., Hoyland, K. & Lia, L., 2023. Review of River Ice Force Calculation Methods. *Cold Regions Science and Technology*, 209, pp. 1–17.

Roeder, C. W., Lehman, D. E. & Thody, R., Fourth Quarter 2009. Composite Action in CFT Components and Connections. *Engineering Journal*, 46, pp. 229–242.

Roeder, C. W., Stephens, M. T. & Lehman, D. E., 2018. Concrete Filled Steel Tubes for Bridge Pier and Foundation Construction. *International Journal of Steel Structures*, 18(1), pp. 39–49.

Rotter, J. M. & Sadowski, A. J., 2017. Full Plastic Resistance of Tubes under Bending and Axial Force: Exact Treatment and Approximations. *Structures*, 10, pp. 30–38.

Sadowski, A. J. et al., 2020. Critical Buckling Strains in Thick Cold-Formed Circular Hollow Sections under Cyclic Loading. *ASCE Journal of Structural Engineering*, 146(9), pp. 1–13.

SeismoSoft, 2010. *Earthquake Engineering Software Solutions – SeismoMatch.* [Online] Available at: https://seismosoft.com/products/seismomatch/. [Accessed 28 April 2023].

SeismoSoft, 2012. *Earthquake Engineering Software Solutions – SeismoArtif.* [Online] Available at: https://seismosoft.com/products/seismoartif/. [Accessed 28 April 2023].

SeismoSoft, 2015. *Earthquake Engineering Software Solutions – SeismoSignal.* [Online] Available at: https://seismosoft.com/products/seismosignal/. [Accessed 17 May 2023].

Shahgholipour, F., Afsari, N. & Ghaseminejad, V., 2024. Correlation between Shear Wave Velocity and Standard Penetration Test for Nowshahr and Chalus, Iran. *Iranian Journal of Geophysics*, 17(6), pp. 37–52.

Sil, A. & Haloi, J., 2017. Empirical Correlations with Standard Penetration Test (SPT)-N for Estimating Shear Wave Velocity Applicable to Any Region. *International Journal of Geosynthetics and Ground Engineering*, 3(22), pp. 1–13.

Silva, P. F. & Manzari, M. T., 2006. Nonlinear Pushover Analysis of Bridge Columns Supported on Full-Moment Connection CISS Piles on Clays. *Earthquake Spectra*, 24(3), pp. 751–774.

Silva, P. F. & Seible, F., 1999. *Design Example of a Multiple Column Bridge Bent under Seismic Loads Using the Alaska Cast-in-Place Steel Shell*, San Diego: University of California.

Silva, P. F. & Srithiran, S., 2011. Seismic Performance of a Concrete Bridge Bent Consisting of Three Steel Shell Columns. *Earthquake Spectra*, 27(1), pp. 107–132.

Silva, P. F., Srithiran, S., Seible, F. & Priestley, M. J. N., 1999. *Full-Scale Test of the Alaska Cast-in-Place Steel Shell Three Column Bridge Bent*, San Diego: University of California.

Singhal, M. K. & Walls, J. C., 1993. *Evaluation of Wind/Tornado-Generated Missile Impact*, Oak Ridge, TN: Oak Ridge National Laboratory.

Stafford, P. J., Mendis, R. & Bommer, J. J., 2008. Dependence of Damping Correction Factors for Response Spectra on Duration and Numbers of Cycles. *ASCE Journal of Structural Engineering*, 134(8), pp. 1364–1373.

Stephens, M. T., Berg, L. M., Lehman, D. E. & Roeder, C. W., 2016. Seismic CFST Column-to-Precast Cap Beam Connections for Accelerated Bridge Construction. *ASCE Journal of Structural Engineering*, 142, pp. 1–14.

Stephens, M. T., Lehman, D. E. & Roeder, C. W., 2015. *Concrete-Filled Tube Bridge Pier Connections for Accelerated Bridge Construction*, Seattle, WA: California Department of Transportation CALTRANS Report CA15-2417.

Stewart, J. P. et al., 2011. Representation of Bidirectional Ground Motions for Design Spectra in Building Codes. *Earthquake Spectra*, 27(3), pp. 927–937.

Tan, Q. et al., 2021. Design-Parameter Study on Seismic Performance of Chevron-Configured SCBF's with Yielding Beams. *Journal of Constructional Steel Research*, 179, pp. 1–21.

Trombetti, T. et al., 2008. Correlations between the Displacement Spectra and the Parameters Characterizing the Magnitude of the Ground Motion. *The 14th World Conference on Earthquake Engineering*, Beijing, China, 12–17 October 2008, pp. 1–8.

Turkish Ministry of Interior, 2020. *AFAD: Turkish Accelerometric Database and Analysis System*. [Online] Available at: https://tadas.afad.gov.tr/login. [Accessed 24 April 2023].

U.S. Nuclear Regulatory Commission, 2007. *NUREG*-0800 – *Standard Review Plan* Section 3.5.3, Rockville, MD: NRC.

United States Atomic Energy Commission, 1974. *Regulatory Guide 1.76 – Design Basis Tornado for Nuclear Power Plants*, Rockville, MD: Directorate of Regulatory Standards.

United States Department of Defense, 2007. *UFC 4-010-01: DoD Minimum Antiterrorism Standards for Buildings*, Washington, DC: US Army Corps of Engineers, Naval Facilities Engineering Command, Air Force Civil Engineer Support Agency.

United States Department of Defense, 2008. *UFC 3-340-02: Structures to Resist the Effects of Accidental Explosions*, Washington, DC: US Army Corps of Engineers – Naval Facilities Engineering Systems Command – Air Force Civil Engineer Support Agency.

United States Geological Survey, 2023. *USGS Earthquake Hazard ToolBox*. [Online] Available at: https://earthquake.usgs.gov/nshmp/. [Accessed 27 April 2023].

US Geological Survey and the California Geological Survey, 2015. *Center for Engineering Strong Motion Data*. [Online] Available at: https://www.strongmotioncenter.org/. [Accessed 24 April 2023].

USDOD, 2023. *UFC 3-301-01: Unified Facilities Criteria (UFC) – Structural Engineering*, Washington, DC: Unites States Army Corps of Engineers.

Whitney, M. W., Harik, I. E., Griffin, J. J. & Allen, D. L., 1996. Barge Collision Design of Highway Bridges. *ASCE Journal of Bridge Engineering*, 1(2), pp. 47–58.

Index

Note: **Bold** page numbers refer to tables and *italic* page numbers refer to figures.

Kolibri 360 GmbH
Ferdinand-Jühlke-Straße 7
99095 Erfurt, Deutschland
produktsicherheit@kolibri360.de